The Institute of Biology's
Studies in Biology no. 106

Aquaculture

P. J. Reay
Ph.D.

Lecturer in Fish Ecology,
School of Environmental Sciences,
Plymouth Polytechnic

Edward Arnold

© P. J. Reay, 1979

First published 1979
by Edward Arnold (Publishers) Limited
41 Bedford Square, London, WC1B 3DQ

ISBN: 0 7131 2721 X

British Library Cataloguing in Publication Data

Reay, P J
   Aquaculture. – (Institute of Biology.   Studies in
   biology; no. 106   ISSN 0537-9024)
   1. Aquaculture
   I.   Title   II.   Series
   639'.3   SH135

ISBN 0-7131-2721-X

Printed and bound in Great Britain at
The Camelot Press Ltd, Southampton

General Preface to the Series

Because it is no longer possible for one textbook to cover the whole field of biology while remaining sufficiently up to date the Institute of Biology has sponsored this series so that teachers and students can learn about significant developments. The enthusiastic acceptance of 'Studies in Biology' shows that the books are providing authoritative views of biological topics.

The features of the series include the attention given to methods, the selected list of books for further reading and, wherever possible, suggestions for practical work.

Readers' comments will be welcomed by the Education Officer of the Institute.

1979   Institute of Biology
                                                        41 Queen's Gate
                                                        London SW7 5HU

Preface

While devotees and sceptics argue about its future significance, aquaculture continues to develop into an important industry in many parts of the world, and has become a very topical field of applied biology. Although its role in the future will depend on the extent to which the biological promise is tempered by economic reality, there is considerable enthusiasm and optimism among many biologists anxious to see a worthwhile application of their fields of interest. To some extent the booklet is written for such enthusiasts, but I also hope to attract biologists who are unaware of the potential of aquaculture as an industry and field of applied research. It should also be of value to those interested in the subject but without biological training. The main theme has been to look at aquaculture in terms of the modification of biological systems. Apart from the commercial application, this can provide fascinating insights into both the biology of aquatic organisms and the process of domestication.

Plymouth, 1978   P. J. R.

Contents

1 Introduction

1.1 Definition and scope

From the biological viewpoint, aquaculture can be regarded as *man's attempt, through inputs of labour and energy, to improve the yield of useful aquatic organisms by deliberate manipulation of their rates of growth, mortality and reproduction.* To the economist it is the production of such organisms *from a basis of site or stock ownership or leasehold*, and certainly the biological manipulations can rarely be justified for common property (open access) resources. Clearly aquaculture is the *aquatic counterpart of agriculture*; the principles and many of the problems are the same, though there are unique characteristics associated with the aquatic medium, and with the entirely different set of organisms involved. The basic aquaculture system is illustrated in Fig. 1–1, in relation to the sections of this book.

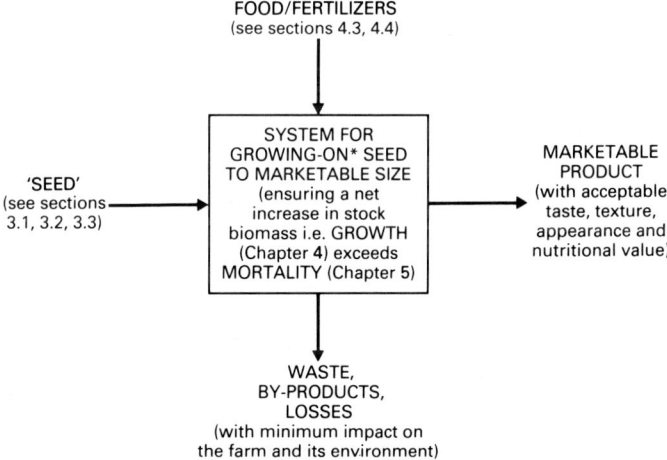

FOOD/FERTILIZERS
(see sections 4.3, 4.4)

'SEED'
(see sections 3.1, 3.2, 3.3)

SYSTEM FOR GROWING-ON* SEED TO MARKETABLE SIZE (ensuring a net increase in stock biomass i.e. GROWTH (Chapter 4) exceeds MORTALITY (Chapter 5)

MARKETABLE PRODUCT
(with acceptable taste, texture, appearance and nutritional value)

WASTE, BY-PRODUCTS, LOSSES
(with minimum impact on the farm and its environment)

Fig. 1–1 Diagram of the basic aquaculture system in relation to its organic inputs and outputs. (Not all systems require an addition of food by man.) (* growing-on = on-growing = fattening.)

Like agriculture, aquaculture has evolved from hunting and gathering. These methods have effectively been replaced by culture methods on land but they still account for about 90% of the world yield of aquatic organisms and are here referred to as *conventional* or *capture fisheries*. In practice, primitive aquaculture may differ very little from capture fishing, but the biological modification in the latter is only through the number of deaths caused by capture (i.e. fishing mortality), and no inputs of labour or energy are made in order to deliberately modify the population processes.

From such considerations, aquaculture can become an uneasy hybrid between fishing and agriculture, and this has given way to a range of different terms to describe the various practices: fish farming, fish husbandry, aquafarming, fish culture, fish cultivation, mariculture, as well as aquaculture itself. *Aquaculture* will be used in this book as an all-embracing term, and *culture* and *farming* as interchangeable synonyms usually with an appropriate prefix (e.g. oyster culture). The term *fish* will often be used loosely to embrace all relevant aquatic animals unless obvious from a particular context. Otherwise, *finfish* (effectively teleosts) will be distinguished from *shellfish* (molluscs and crustaceans) and other smaller groups. *Fish farming* is best regarded as a colloquial alternative to aquaculture though it would not cover the culture of plants, and in a restricted sense would refer only to finfish.

Most aquaculture takes place to provide food for direct human consumption, and this must be seen as its dominant future function in an increasingly hungry or greedy world. In addition, indirect human consumption is achieved by producing organisms for animal (including fish) feed, bait or sport. Several of the most valuable aquacultural products, such as pearls and ornamental fish are not for food at all, and some cultured fish are used for biological control; for example, grass carp* and the mosquito fish (*Gambusia affinis*). As well as this range of product uses, farms are of additional value in providing employment and utilizing waste.

A recent survey suggested that 314 species of finfish, 74 crustaceans, 69 molluscs, 43 algae, 13 angiosperms (including rice), 12 sponges, 9 amphibians, 4 reptiles, 3 rotifers, 2 annelids, 2 mammals and 1 echinoderm are the subjects of aquaculture. This list excludes purely or primarily aquarium species and those only cultured in laboratories. In contrast to agriculture there are many more animals than plants; they are from a greater range of taxonomic groups; they are poikilothermic, often carnivorous and typically highly fecund with a free larval stage in the life-cycle; and there is very limited domestication – only common carp and rainbow trout are considered to have domesticated varieties at present. Both species are very important in world aquaculture and have been successfully introduced by man to areas outside their natural range of distribution. This also applies to cupped oysters (especially the Pacific oyster) and tilapias (especially *Sarotherodon mossambica*), but other important species such as milkfish and yellowtail are cultured only in a limited area. There are also some very specialized animals, both in a

---

* Scientific names not given in the text appear in the Appendix with further details of the cultured species.

biological and a marketing sense, such as the mud-skipper which is cultured on a small scale in Taiwan. (CHEN, 1976.) Some of the species used in modern aquaculture are illustrated in Fig. 1–2.

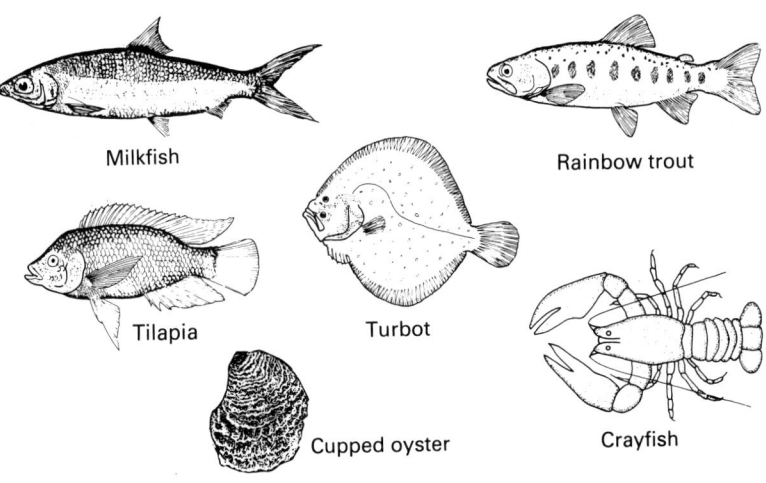

Fig. 1–2   Some of the species used in aquaculture. (Drawings not to scale.)

The simplest types of aquaculture, involving the minimum of manipulation and hence the least departure from nature, are usually described as *extensive*, *semi-culture* or *open* systems. They would include the transplantation of oysters to favourable growing areas, and the addition of fertilizers to natural ponds. At the other extreme would be a very *intensive*, *closed* system, equivalent to a broiler-house unit on land and involving almost complete control over the organism and its environment. The term intensive refers to the high density and production of organisms per unit area compared to that occurring naturally. Aquaculture systems can be classified in a number of different ways, and Table 1 shows a simple division into eight main types based on the presence or absence of: (*a*) hatchery production of young stages (seed), (*b*) holding facilities, and (*c*) intensive feeding by the farmer. Particular attention should be paid to the difference between Types V and VIII. Ranching or stocking involves the release of hatchery-reared juveniles into the natural environment, and is an important alternative to growing-on in holding facilities.

1.2   Status, history and potential

According to world statistics collected by the Food and Agriculture Organisation (FAO) of the United Nations, 6 029 289 tonnes were produced through aquaculture in 1975. This was about 10% of total

Table 1   The main types of aquaculture, a classification based on degree of biological manipulation.

Type	Seed supply from hatcheries	Holding facility provided	Stock fed by farmer	Examples
I	No	No	No	Bottom culture of oysters and mussels
II	No	Yes	No	Raft or tray culture of oysters and mussels, milkfish culture in ponds and pens
III	No	No	Yes	? (unlikely)
IV	No	Yes	Yes	Yellowtail culture, eel culture
V	Yes	No	No	Salmon ranching, sturgeon stocking, crayfish stocking
VI	Yes	Yes	No	Nori (seaweed) culture, some oyster culture, some pond culture of finfish
VII	Yes	No	Yes	(Trout experiments in Scottish lochs)
VIII	Yes	Yes	Yes	Most trout and salmon culture, Japanese prawn culture

Notes:

(i) A holding facility is any constructed structure, such as a raft, tank, cage or artificial pond for holding and restricting stock movements (further details in §2.3).

(ii) Type I would also include capture fisheries, and simple aquaculture manipulations associated with them, such as predator control and habitat improvement. Otherwise it is only possible with sessile bivalves.

(iii) Types III and VII are only included for completeness. A combination of unrestricted movement (which would apply to competitors and predators as well as stock) and costly artificial feeding is unlikely, though some work has been done on the use of feeding stations to which the fish are 'called' or attracted with a sound stimulus.

aquatic production in the same year, the difference, of course, being made up by capture fisheries. Aquaculture production is composed by weight of 66.0% finfish (mostly freshwater), 16.2% molluscs, 0.3% crustaceans and 17.5% seaweeds. Some national figures are given in Table 2, from which it is clear that countries in Asia and the Far East are responsible for most of the world production.

One reason why aquaculture is so prominent in the East is that it almost certainly originated there. The earliest records come from China where it is thought that carp culture was being practised 4000 years ago, although the first treatise on the subject did not appear until 475 B.C. (HICKLING, 1971). Fish culture was probably introduced to England from

Table 2   Aquaculture production, by country, in 1975: (a) in 10^3 tonnes, (b) as a percentage of the country's total aquatic production, (c) the main groups of organisms, in order of importance. (Based on data from *FAO Yearbook of Fishery Statistics* (1975) and PILLAY, T. V. R. (1976) *FAO Technical Conference on Aquaculture*, Paper R. 36.)

	a	b	c	
				Countries producing 10 000–100 000 t
China	2500	37.0	F S	Bangladesh, Nigeria, Poland, Mexico, Italy, Germany (Fed. Rep.), Vietnam (Rep.), Malaysia, Yugoslavia, Romania, Hungary, Madagascar, Germany (Dem. Rep.), Czechoslovakia, Israel, Denmark, Brazil
Japan	945	9.0	S M F C	
India	494	21.2	F C	
Korea (Rep. of)	300	14.2	S M	
U.S.S.R.	210	2.1	F	
Spain	162	10.6	M	*Countries producing 1000–10 000 t*
U.S.A.	151	3.2	M F	Australia, Sri Lanka, Egypt, Canada, Zaire, U.K., Cuba, Hong Kong, Norway, Chile, Austria, Finland, Belgium, Tanzania, Burma, El Salvador
Indonesia	144	10.4	F C	
Taiwan	127	16.4	F M S C	
Philippines	125	9.4	F M C S	*Countries producing 100–1000 t*
Thailand	106	7.7	F M C	Ecuador, Greece, New Zealand, Singapore, Uganda, Kenya, Nepal, Venezuela, Switzerland, Ireland, Senegal
France	104	12.9	M F	
Netherlands	102	29.2	M	

F – (teleost) fish, M – (bivalve) molluscs, C – (decapod) crustacea, S – seaweeds.

Europe, when carp were brought into the country for the first time in the fifteenth century. Prior to this, fish were kept in ponds to supply food at times when fresh meat was scarce. When the development of sea fisheries and improved methods of preservation and transport in the nineteenth century brought fresh seafish in quantity to all parts of the country, most of the ponds and culture techniques were abandoned. Marine aquaculture is of much more recent origin in the U.K., but has a history of at least 600 years in France (mussels) and 300 years in Japan (oysters and seaweeds) and Taiwan (milkfish). It started here in the nineteenth century when the hatchery techniques used to produce young trout and salmon for stocking rivers and lakes in Europe and North America, were extended to other fish species including commercially important seafish such as plaice (*Pleuronectes platessa*). Overfishing was seen to be a problem in some areas at the end of the nineteenth century, and the release of large numbers of *early* post-larvae from hatcheries into the sea was mistakenly considered an effective way of boosting depleted wild stocks.

Although the marine hatcheries closed down in the early years of this century, the work has been vigorously resumed in the last twenty years – this time mainly in connection with tank and cage culture. Modern aquaculture is a development primarily of the post-war period and the development, both in terms of technical evolution and expanding production, has been very rapid.

In view of its recent history, aquaculture would appear to have an important future in a world where more food is needed, yet where production from both agriculture and fisheries appears to be approaching its limits. The exact potential for aquaculture is of course not known, but the FAO estimate of a five-fold increase over the next thirty years is reasonable provided that certain constraints to development can be removed. Not all constraints have a biological origin; they include the availability of sites, water, finance, trained manpower, feeds, fertilizers, seed and suitable markets, and in many cases the constraints are legal or administrative.

Increased production will come about by increasing the intensity of production per unit area, and by increasing the areas under cultivation. About 4 million tonnes of fish a year are currently produced in Asia and the Far East from 2 million hectares. It is estimated that 22 million hectares are available, and that with improved techniques these could yield 120 million tonnes each year. Similarly, there are thought to be 440 million hectares of coastal wetlands in the world which are mostly tropical mangrove swamps. If only 10% of this area could be used, then at least 100 million tonnes of extra fish could be produced each year, but the ecological impact of swamp destruction first needs to be carefully considered.

On both a national and a global scale aquaculture will become an increasingly important component of aquatic production. Even though the latter is less than 3% of the world harvest from all organisms, it is, in some localities and in certain countries, of great importance, particularly as many aquatic products are of high market value.

1.3   Intensities of production

Intensities of production usually refer to the rate of fish produced per unit area, for example, $kg\ ha^{-1}\ yr^{-1}$ or $kg\ m^{-2}\ day^{-1}$, and the more intensive the system (§ 1.1) the higher such production is likely to be. Implicit here is some concept of efficiency, and if land or water area is seen to be a limiting resource in aquaculture development, then intensive production will be considered efficient. Yet it will probably be achieved only by increasing inputs of capital, labour, energy, water volume or feed, and any of these could also be regarded as limiting factors. In some situations, therefore, it can be more appropriate to consider as efficient that system which produces the most fish, or protein, per weight of feed, per man-hour, or at lowest GER (gross energy requirement). Specifically one could compare a Thai carp/tilapia subsistence unit which requires less than 1 GJ (gigajoule) of energy input, more than 1 ha of land and 120 man-days to produce 1 t of fish, with U.K. trout ponds which require 50–100 GJ, 0.01–0.1 ha and 25–50 man-days to produce the same quantity (EDWARDSON, 1976). So to minimize confusion, reference should always be made to inputs and type of system when comparing intensities of production and efficiency.

The highest recorded levels of fish production from static unaerated pond water are 49.9 kg ha^{-1} day^{-1} over a 120-day period using a mixture of species (polyculture) in 400 m^2 ponds in Israel. The fish were fed with protein-rich pellets but substituting the feed with cow manure still gave as much as 31.7 kg ha^{-1} day^{-1} (MOAV et al., 1977). These and other reports suggest that over 10 t ha^{-1} yr^{-1} can be realized from static ponds, particularly at low latitudes where the growing season is long. Many existing ponds produce less than a tenth of this, but in China mean production in different areas is 3–4 t ha^{-1} yr^{-1} (ANON, 1977). Much higher production is possible if aeration or water renewal is available. An extreme example is the 750 t ha^{-1} yr^{-1} of edible tissue obtainable from the Spanish mussel rafts in tidal waters – a system of aquaculture that also requires no supplementary feeding unlike intensive finfish systems.

Production intensity figures do have some value for comparative purposes especially where land/water area is a scarce resource, but the need to consider other inputs has been stressed, and the danger of extrapolating from small areas and time periods should also be recognized. A further general point is that some published figures refer to gross production and others to net production (i.e. taking into account the initial weight of organisms). This initial weight is normally small, but the two types of production should be distinguished where possible.

Yield is a more appropriate term than production since in the ecological sense, the latter would also include animals which have died during the growing period. The farmer should be interested in minimizing the gap between production and yield, but the latter will be his criterion of success.

1.4   The biological basis

Plants account for about 85% of the terrestrial food harvest whereas in water the equivalent figure is less than 2% (rice is included in *agri*cultural statistics). Most of the harvested aquatic plants are benthic and are therefore restricted to shallow water which in the sea forms only a narrow coastal band. The plants involved are either macroalgae (seaweeds) or angiosperms (in fresh water), but phytoplankton is often the dominant plant form in most bodies of water. It however, consists of unicellular algae (diatoms, flagellates, dinoflagellates, desmids, etc.) whose small size, amongst other characteristics, hinders direct exploitation by man.

Although some finfish eat plants, aquatic herbivores feeding on the phytoplankton are typically either filter-feeding bivalves or the small animals, mostly crustaceans, of the zooplankton. The former are readily exploited though mostly restricted to shallow water, whereas the latter, comprising the bulk of aquatic herbivore biomass, are not. The zooplankton, however, provide the trophic link between phytoplankton and many species of commercially important fish.

Thus man tends to exploit aquatic ecosystems through animal species at primary, secondary or even tertiary carnivore levels. But even at these levels only certain species are of commercial significance and they are selected because they are abundant, available to boats and gear, easy to handle and process, and acceptable to the consumer.

Estimates of the potential yield that is available from the oceans vary widely from less than twice to forty times the present level of about 60×10^6 t yr^{-1}, depending to a large extent on whether resources like Antarctic krill (*Euphausiacea*), oceanic squid (mostly *Ommastrephidae*), and mesopelagic lanternfish (*Myctophidae*), at present only lightly exploited, have an important role to play in the future. Whatever the viewpoint chosen, however, there is clearly some limit to the total catch, and various constraints can be identified.

1   *Non-biological* (social, economic, political, technological) – mainly determine which species and stocks can be exploited and the efficiency of exploitation.

2   *Biological* – determine how much can be taken.
   (i)   Constraints, imposed by the ecosystem dynamics, to the abundance/production of a stock:
      (*a*) rate of primary production
      (*b*) trophic level of stock
      (*c*) ecological efficiency of energy transfer between trophic levels
      (*d*) abundance/production of competitors and predators.
   (ii) Constraints, imposed by the population dynamics, to the yield sustainable from the stock:
      (*a*) relationship between parent stock and recruitment
      (*b*) relationship between growth and mortality.

The first set of biological constraints are imposed by climatic and hydrographic factors affecting primary production and also by the structure of the community. In a wild situation, the only way of avoiding these constraints and thus increasing potential yield, is through species selection. Whilst bearing in mind the non-biological factors, it would be a wise policy in future to expand the range of species exploited at any one trophic level and also to exploit at as low a trophic level as possible. The amount that can be removed (the yield) from a stock, however, does not depend entirely on its abundance or production. It is the population dynamics of the stock which will also determine the maximum sustainable yield. Further details are given by CUSHING (1977). The important point to note here is that the only control is through the number of fish deaths and the age at which the fish begin to be harvested.

In practical terms this means the numbers of boats, their size, time spent fishing, areas fished, and the selectivity of the gear.

In conventional capture fishery situations, therefore, the only ways in which the yield can be controlled is through choice of species and fishing strategy. By contrast, in aquaculture, an attempt is made to remove the constraints to production and yield by manipulating the appropriate ecosystem and population processes. Thus, in addition to the appropriate selection of a species, some attempt is made to increase primary production, increase efficiency, decrease food-chain length and remove competitors by the appropriate manipulation – either of the stock directly or through its environment. At the population level, growth is encouraged, recruitment is maximized but becomes independent of the fish to be harvested, and natural mortality is minimized. In these ways the normal limits to yield are lifted, there is no need to conserve a large spawning stock, and fishing mortality will approach 100%.

As well as attempting to remove the various constraints to production and yield, predictability is increased and the risk of doing badly is decreased if aquaculture techniques are used. This is achieved by simplifying the natural systems in terms of the diversity of species number, age group composition and environmental variability. Such systems are inherently unstable, however, and there is a risk of doing *very* badly if the manipulations (e.g. feeding, disease control, aeration) which support the artificial system are not sustained. This is because homeostatic regulatory mechanisms will tend to restore the system to its natural state, usually through mortality. The more intensive the system the greater the inherent instability.

1.5　Interaction with other industries

So far, aquaculture has been considered in isolation, but of course as an operating industry it inevitably interacts with other human activities such as agriculture, capture fisheries, water storage, waste disposal, shipping, electricity generation, recreation and amenity. The interactions may be *competitive* or *integrative*, and the former may place severe constraints on aquaculture development. Most aquaculture systems require some input of scarce resources (capital, land, water, feed, energy, etc.) and also impinge visually on the landscape. Many, while very sensitive to the polluting activities of other water-users, cause organic pollution themselves. (Some of the potentially deleterious effects of coastal aquaculture are reviewed by ODUM, 1974.)

These points could be used as arguments to inhibit the development of aquaculture, but it is more valuable to consider how the various aquaculture systems can integrate with existing industries and capitalize on the many beneficial aspects. An example here is the ability of many systems to utilize waste (§§ 4.4, 4.5). A *waste* is a resource where the cost and complexity of using it appear to be greater than the returns from using it, and aquaculture is able to remove the 'waste label' from a whole range of resources and produce valuable protein and other products at the same time.

Pig and duck farming is often carried out in conjunction with a fish pond, and here the fish benefit either directly or indirectly from the faeces and uneaten foods of the other animals. The pond is a feature of the integrated rural development units of China and is now being advocated in other parts of the world. Wastes, from night soil to plant refuse are turned into readily available and much-needed protein in the pond, and it is only an extension of this to consider growing fish in the waste water from sewage treatment plants.

Water is often a scarce resource in agriculture but use can be made of irrigation water for fish production, thus increasing overall yields. The most well-developed system of this type is to grow fish in rice fields (*rice-cum-fish culture*), which gives a valuable crop in addition to the rice. It is practised in parts of Asia and North America and 35 species of fish have been involved. In many areas the practice has died out because of the use of short-stemmed rice varieties (needing less water) and chemical pesticides, but there is a general revival of interest.

There is similar scope for integration between aquaculture and capture fisheries, for example in *sea-ranching* and *stocking*, the production of seedlings in hatcheries are released into the wild to boost natural stocks and benefit fishermen. An extension of this would be the genetic improvement of wild stocks in view of the possibility that, by selectively removing the fastest growing fish, fishing itself may have caused a decline in growth potential.

Another area for integration is in the provision of *bait fish*. Obtaining adequate supplies of suitable bait for pole and line fisheries is a real constraint to realizing the potential of stocks of tuna, like the skipjack (*Katsuwonas pelamis*), in parts of the Pacific, and investigations are proceeding on the possibilities of culturing sharpnose mollies (*Poecilia vittata*) and small tilapias for this purpose. Some of the milkfish cultured in Taiwan are harvested early and used as bait, and there are also well-established farms for bait-minnows (e.g. the golden shiner, *Notemigonus crysoleucas*) in North America. A technique has recently been developed in Florida for the culture of lugworms (*Arenicola cristata*), a favourite bait for sea-anglers.

2 Selecting a System

2.1 Species selection

In theory almost any species, if considered useful enough, can become the object of commercial exploitation, and also a potential candidate for aquaculture. Although the initial choice may seem wide, it is restricted in practice by the biological, technological, economic and marketing criteria that determine commercial viability (WEBBER and RIORDAN, 1976). Relatively few species can be farmed on a large-scale at the present time, and for a specific site the choice is even more restricted.

Marketability is usually the most critical consideration, and species characteristics which may affect this are taste, appearance, texture and ease of processing. None of these are immutable and can to some extent be adjusted to meet the consumer's demands. Social and cultural factors are often more significant even than economic ones, and nutritional value is rarely a primary concern of the consumer. It is generally difficult to introduce 'new' products, and the conservative tradition with regard to shellfish in the U.K. (compared to many other European countries), is the major constraint to the expansion of oyster culture even after years of market research and promotion of the industry by the White Fish Authority (WFA) and others.

To some extent demand can be stimulated by lowering the price, but in the long run the product must command an appropriate selling price in relation to production costs, and it is these costs which affect economic viability. Biological characteristics such as growth-rate, conversion efficiency, seed availability and hardiness (tolerance to crowding, abiotic factors and pathogens) will be important considerations influencing production costs and hence species choice. However, desirable biological characteristics are of limited value if they can only be realized at high cost, or are associated with low demand, and it is reasonable to suppose that biological and technological criteria will often be secondary to marketing and economic ones.

Biological screening is nevertheless an important part of species selection. In the U.K. the turbot has been selected by the WFA, Ministry of Agriculture, Fisheries and Food (MAFF) and several industrial companies as the main contender for the first farmed marine teleost (JONES, 1972). This was after earlier work had been carried out on other flat-fish, notably plaice. The advantages that turbot has over plaice are higher market value, greater tolerance to crowding, faster growth, and cheaper feeding costs. Its original disadvantage was unreliability of seed production, but the high market value has stimulated the research appropriate to overcoming this problem, and it has now largely been solved (§ 3.3). The realistic choice of species in the U.K. is shown in Table 3. In other parts of the world, particularly at lower latitudes, the choice is wider because of less inhibited consumer acceptance, and because of the inherent higher species diversity associated with tropical and sub-tropical regions.

Table 3   Choice of species for food in the U.K.

Brown trout	Rainbow trout	Atlantic salmon	American brook charr	Coho salmon	Common eel	Common carp	Turbot	Dover sole	Lobster	Signal crayfish	Pacific oyster	European flat oyster	Great scallop	Mussel	
x	x	x	x	?	x	x	x	?	?	?	x	x	x	x	Commercially farmed in U.K. at present
	x		x	x		x				x	x				Not indigenous to U.K.
x	x	x	x	x		x				x					Early life history requires fresh water
x	x		x		x	x				x					Grown-on in fresh water
	x	x		x	x		x	x	x		x	x	x	x	Grown-on in sea-water
x	x	x	x				x	x			x	x	x		Seed now available from hatcheries or breeding ponds
									x				x	x	Seed at present only available from wild
						x				x	x	x	x	x	Food costs cheap or nil
					x		x	x	x						Heated water needed for economic farming
x	x	x	?	?			x	x	x	?		x	x		Market demand high in U.K.

More than one species may be cultured at one site, and particularly if these are kept in the same holding facility at the same time, the technique is known as *polyculture*. The high yields from Israeli and Indian static ponds are achieved by keeping a mixture of species in the same pond. Further experiments on the species used and their stocking ratios are being undertaken, but in recent experiments in India, silver carp, common carp, grass carp, four species of Indian major carp (e.g. rohu) and grey mullet were kept together. In Israel a combination of silver carp, common carp, tilapia, and grey mullet is commonly used.

In the wild, of course, several species of fish live together as a community. Typically, most aquaculture systems will simplify this community to an extreme by utilizing one species (*monoculture*) and eliminating the rest which may include predators and competitors. It is now recognized that one species cannot often make optimal use of an aquaculture system. Several species, each with a different feeding habit, can most efficiently utilize the natural food in a pond. Although most suitable for pond situations where natural production is an important food source, polyculture has also been considered for raft culture of bivalves (oysters near the surface, scallops in deeper waters) and in sea-cage and freshwater salmonid culture with lobsters or crayfish as bottom scavengers eating faeces, corpses, and uneaten food. Sometimes a second species is introduced for a more specific purpose, such as to control the numbers of young produced by the prime species (e.g. Nile perch, *Lates nilotica* with tilapia species).

Species selection for polyculture has to ensure that all the species used should not only have individually desirable characteristics, but should also be compatible and, if possible, mutually beneficial. Many combinations would clearly not work. Among the less obvious are those invertebrate–vertebrate couplings which might enable completion of a parasite life-cycle that would be impossible under monoculture. Another example is the incompatibility of either grey mullet or tilapia with milkfish. The latter species depends for its food on a bed of benthic algae (§4.3) which the others would destroy through foraging and, in the case of tilapia, nest-building.

Until the techniques of aquaculture become more advanced, and in particular, until it becomes possible to produce intraspecific varieties designed for particular situations, there will continue to be advantages in introducing from one country to another the few well-established candidate species. World markets exist for some of these species and relatively few bio-technical problems exist. Nevertheless, national and international legislation is now increasingly hostile to the necessary free movements of organisms, the main reason being the need to control the introduction of pathogens (§ 5.5). But the introduced species themselves can have a deleterious effect on native ecosystems if they escape from the confines of the farms, particularly if they are able to breed in their new environment and establish self-sustaining populations.

Fortunately, perhaps, the two main (and very successful) introductions to the U.K. – rainbow trout from North America, and Pacific oysters from Japan – only rarely breed in the wild. Some introductions, however, are made in order to establish self-sustaining populations, and the signal crayfish in Europe is an example. There are many instances where introduced species have caused serious problems. These are reviewed by ROSENTHAL (1976) who also suggests a code of practice to be followed when future introductions are contemplated.

2.2　Site selection and water quality

The most important site characteristics from the biological point of view are those related to the water supply. Water is the medium surrounding the fish, and the resulting interface includes the sites of gaseous exchange, ionic balance, removal of waste products, and the determination of body temperature. For bivalves and others the medium also contains food in suspension. The quantity of water available will determine the holding capacity and production of an aquaculture system, and the quality of the water will in addition influence the choice of species, holding facilities and type of farm operation.

In theory, any deficiency in water quality can be rectified by *pre-treatment* processes, and if there is too little water, then *reconditioning* and re-use can be considered. As aquaculture intensifies, these methods are becoming increasingly important but tend to be expensive and the choice sites for aquaculture are still those with an abundant supply of suitable water.

The amount of water in a static pond will determine the biomass of fish able to be held per unit volume or area, and thus the intensity of production. In an unmodified system the first limiting factor will probably be food supply, but in ponds where artificial feeding or fertilization is used then the effective limiting factor will usually be *oxygen depletion* or *self-pollution* with waste products, in particular ammonia. If these constraints are removed, for example by frequent exchange of water or artificial aeration, then the factor determining holding density will be the mutual tolerance of the individual fish. Most teleosts and molluscs are able to live in very close proximity as long as they have adequate food, oxygen and waste removal. The example of successfully growing carp in tanks at 0.2 kg l^{-1} (a fish:water volumetric ratio of 1:2 compared with 1:20 000 in a static carp pond) will suffice to show this. Several species of decapod crustacea such as lobsters, crayfish and caridean prawns, however, are prone to cannibalism unless kept separately or provided with a surplus of refuges.

In continuously flowing water situations the *rate of volume flow* becomes the decisive factor. In fact the use of flowing water is an alternative to *aeration* or *oxygenation* when attempting to remove the oxygen/waste constraint to holding density levels. If both are used, then even better results are possible. At one trout farm in the U.K. a production of 90 t yr^{-1} is possible from a daily flow of only 5 million litres because the tanks are oxygenated. The water requirement is only a tenth of that which would be needed without oxygenation. (A minimum level of 5 mg O_2 l^{-1} is needed by trout.)

The use of flowing water can involve directing it through ponds and raceways and in this way some control over flow rate is possible. The alternative is to suspend the stock in the water in cages for teleosts, or in various forms of suspended culture for bivalves, and is most widely practised in the sea where water movement is generated by tides and currents. It is usually the *minimum flow-rates* likely to be experienced at a site that are the most important in site selection. Careful consideration is therefore needed of summer flow-rates in fresh water and periods of slack water in the sea as part of preliminary hydrographic surveys, since these are the times when oxygen problems may arise. However, it is sometimes possible to compensate using either aeration (surface agitation, submerged diffuser systems or flow-induced venturi-type air/water mixers) or oxygenation (the dissolution of pure oxygen or oxygen-enriched air), and if this is done, it enables maximum holding density to be realized.

Temperature is perhaps the most important water quality characteristic because of its direct effect on activity and metabolic processes, its indirect effect through dissolved oxygen level, and the high cost of water temperature manipulation. Species differ in their temperature requirements, and so this factor is important in selecting suitable sites for a predetermined species, or species for a predetermined site. All species have maximum and minimum lethal temperatures (§ 5.2) which have to be avoided, but of equal importance is the need to provide the optimum temperature for growth or conversion efficiency (§ 4.1). In temperate waters the usual problem is that temperatures are too low, particularly during the winter months and this restricts the growing season. When problems with high temperatures do occur, it is sometimes through effects on dissolved oxygen levels. The most economical way of increasing water temperature is to use heated effluent from industry, particularly electricity generation (§ 4.5).

Freshwater and marine sites are normally distinct, but since marine aquaculture is currently concentrated in estuaries and other sheltered coastal waters, *intermediate* or *fluctuating salinities* are often features of marine sites. This is sometimes an advantage in salmonid culture, since it facilitates transfer from freshwater hatcheries to full strength sea-water. The euryhalinity of rainbow trout is currently being exploited in Scotland where the major expansion of farming is in the marine sector. The advantages of using marine sites for trout are that (a) more are available, adequate freshwater sites being scarce, (b) growth is faster, (c) there is a more stable chemical environment, and (d) water temperature is more stable and higher in winter. Other important cultured species that can be grown in either fresh or salt water are tilapias, grey mullet and milkfish.

Contamination of the water source with high concentrations of *heavy metals*, *biocides* or *oil*, can be a serious constraint to the suitability of a site, particularly as these pollutants are difficult to control. *Organic pollution*, however, as long as it is accompanied by adequate aeration, need not be a problem and can be used to advantage since the organic matter can be a direct or indirect food for the stock and a fertilizer. (§§ 4.3, 4.4.) Possible constraints on *waste-water fish culture* are (a) unsatisfactory dissolved oxygen levels, (b) presence of toxic materials with the waste, (c) presence of pathogens, (d) presence of unacceptable tastes and odours, (e) various problems of public health and public acceptance. These problems mainly apply to the use of sewage waste water which may be either pre-treated, diluted, or not treated at all. Even salmonids can be grown satisfactorily in waste water, and the main constraint is probably consumer acceptance.

Bivalves can pose serious health problems in polluted waters both in relation to heavy metals and sewage bacteria. *Depuration* with ozone or ultraviolet light may be necessary prior to marketing.

Post-treatment of the water may be used to ensure an adequate standard of effluent being released back into the natural environment, and without it the pollution laws relating to the site may be regarded as a constraint to its development. It is a sobering thought that a trout farm with a stock of 35 t has a potential stream influence equivalent to that of a village of 1000 inhabitants. The other use of post-treatment is that by reconditioning the water, some at least is able to be recycled and used again. This can give a much greater production from the initial volume of water and can turn an intrinsically unsuitable site into a commercial proposition.

Reconditioning systems can be classified into three types; simple, complex open and complex closed. Even with simple aeration 80% re-use may be possible, but with more complex treatment processes such as sedimentation and filtration as well, completely closed systems with almost 100% re-use are possible. It may in addition be feasible to heat the water in reconditioning systems and conserve the heat along with the water. In this way it may be possible to culture tropical prawns in temperate latitudes (WICKINS, 1976).

The main basis of post-treatment in either open or closed systems is to pass waste water through a biological filter (either percolating or activated sludge) in order to convert ammonia through nitrite to nitrate using nitrifying bacteria. In a closed circulation system the volume of water in the purification tanks at any one time is likely to be 6–8 times that in the holding tanks, but it has been suggested that free-floating nitrifying microcolonies could replace the conventional bulky filtration system with its plastic or gravel media base. As an alternative to biological filtration it may be possible to use ion exchange resins.

2.3   Holding facilities

The holding facility maintains in a physical sense the artificial ecosystem that is essential to most forms of aquaculture. More specifically, its purpose is to enhance yield by: (a) facilitating manipulations of stock and environment; (b) preventing escape of stock and loss of desirable elements (e.g. heat, food organisms); (c) preventing entry of undesirable elements (e.g. predators, competitors, pollutants); (d) facilitating maintenance and harvesting. The type of holding facility used in a particular situation, therefore, will depend on these considerations and will involve more than just biological criteria.

The types of holding facility available for *sessile organisms* (bivalves and seaweeds) are: (a) bottom culture, sometimes protected by enclosure or netting cover; (b) attached to sticks or poles; (c) attached to ropes; (d) contained in bags; (e) contained in trays; (f) attached individually to plastic holders. Ropes, bags, or trays may be on trestles, or suspended from rafts or buoyed long lines (Fig. 2–1). Normally bivalves are grown in

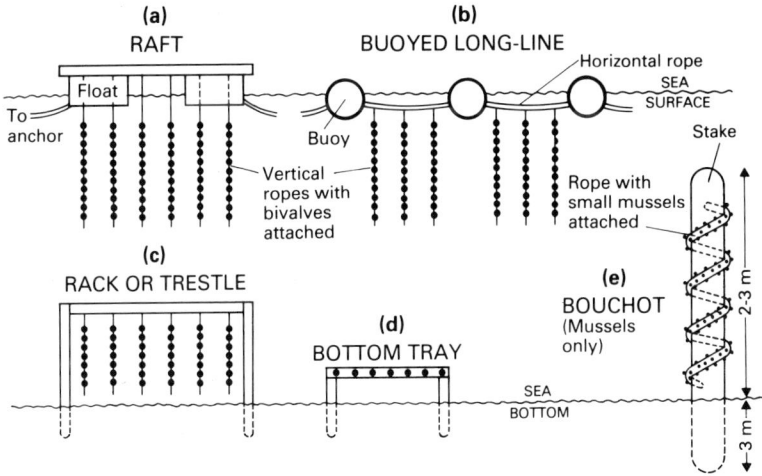

Fig. 2–1　Some of the holding facilities used in bivalve culture; (c), (d) and (e) would normally be used intertidally. (Not to scale.)

the natural environment but tank culture (University of Delaware closed system) and pond culture (French 'claires' for oysters) are also possible.

All methods of holding bivalves are in current use, but there is a trend towards off-bottom culture with the animals suspended either from *rafts* or *long lines*. This generally makes harvesting easier and frees site location from any restrictions imposed by depth or type of bottom. In most cases it also results in fewer problems with predators (particularly crabs) and some parasites (e.g. the copepod, *Mytilicola intestinalis* in the common mussel). Further, because more of the water column can be utilized, production per unit area will normally be higher. Within suspended culture, buoyed long lines seem to be increasingly favoured in Japan at least because these are more resistant to exposed conditions and to some extent free aquaculture from its traditional association with sheltered inshore areas.

For *motile organisms* (fish and crustacea) the following holding facilities are available: (*a*) none (ranching); (*b*) earth ponds; (*c*) lined ponds; (*d*) concrete raceways; (*e*) plastic or fibreglass tanks; (*f*) enclosures; (*g*) cages. (Fig. 2–2). All are in current usage. *Earth ponds* are the most common on a global scale and 82% are fed by fresh water, the rest by tides. Earth ponds are cheap and easy to construct if the ground is soft. Even if sandy and porous, it can be made watertight with clay and dung, as practised along the Israeli coastal strip. The other advantage of earth ponds is that the pond soil facilitates the growth of natural food organisms for the stock (§ 4.3). But the very 'natural' nature of the ponds may bring with it problems of pest (e.g. *Scylla serrata* a burrowing crab which destroys the dykes of milkfish ponds, but is also a secondary crop), parasite (the agent of

Fig. 2–2 Two types of holding facility commonly used for growing salmonids: raceways (left); sea-cages (right).

whirling disease in European trout ponds § 5.3), and competitors and predators which may be difficult to eradicate. Again, low pH caused by the leaching of iron sulphate salts in the subsoil of the dykes may be a problem encountered in some situations, and adequate facilities always need to be available for draining ponds as well as filling them up.

As long as the stock is being fed artificially and is not dependent on natural production, rubber or plastic-lined earth ponds provide an adequate holding facility for several species of fish. More expensive are concrete *raceways*, or *tanks* of a variety of shapes, sizes, and materials widely used in intensive trout rearing in Europe and North America. The decision on which facility to use will almost always be based on economic or engineering criteria in that the fish themselves are only indirectly dependent on the facility, but directly dependent on the aquatic environment created. Thus the availability of oxygen and removal of waste products already described as dominant water quality criteria, will be influenced more by the shape of the container and the pattern of water circulation within it, than by the materials used in construction.

Cage culture of fish is an alternative to the above facilities and ideally involves placing the cage in flowing water, which replenishes the supply of oxygen and removes waste products. Most cages are suspended at the surface on floats which makes for easy access for feeding, maintenance, and harvesting, but such rafts, as in bivalve culture, are sensitive to exposed conditions. To overcome this limitation, the Japanese have developed ikesu cages which can be lowered to, or maintained at, depths out of reach of rough surface conditions, and are equipped with a long feeding funnel.

In all cage culture the mesh size is critical. It must contain the stock and prevent entry by predators, but it must also be large enough for an adequate interchange of water. Mesh size criteria will also apply to bivalve culture where the stock is kept in trays or bags, and here, as in cage fish culture, fouling of the netting by macroalgae and invertebrates will be a major problem. Fouling makes the net heavier, increases drag, and reduces water exchange. It has sometimes been necessary in Japanese waters in summer to change the netting every 2–3 weeks because of fouling. The expense and inconvenience of this can to some extent be reduced by appropriate choice of netting materials, and MILNE (1975) has carried out trials to establish those least susceptible to colonization. Galvanized weldmesh is better (but more expensive) than synthetic fibre which in turn is better than natural fibre. Impregnation with anti-fouling solutions, including copper, are of some help, and it may be possible to exploit seasonal and distributional variation in the intensity of fouling. Thus fouling in brackish water is typically less than in sea-water. The use of biological control in the form of herbivorous siganid fish has been investigated experimentally in the Palau islands and seems to hold some potential.

The final type of holding facility is an *enclosed system* where a natural area is netted or fenced off. It is a cheap alternative to cage culture, particularly in marine areas, and is used by some Norwegian salmon farmers. Environmental control and harvesting are more difficult than in cages, but as a less intensive system, it gives the fish more opportunity to forage on natural foods. More elaborate are the fish pens for milkfish in Laguna de Bay, Philippines. These are areas away from the shoreline and therefore fenced on all sides.

As an alternative to providing a specially constructed holding facility, there is some potential in utilizing existing structures such as marinas, offshore platforms and even vessels.

More information on both site selection and holding facilities will be found in HUET (1973) and MILNE (1972). The former deals mainly with freshwater the latter with marine situations. KORRINGA (1976) provides technical details for bivalve and seaweed culture.

3 Obtaining Seed

3.1 Obtaining seed from wild populations

The ideal aquaculture system is independent of wild fish, as in modern rainbow trout farming. However, because of the difficulties of controlling reproduction (§ 3.2), the culture of many species still remains dependent on wild populations. Some major examples are milkfish, yellowtail, grey mullet, eel, mussels, most penaeid prawns and some oysters.

There are several disadvantages inherent in the harvesting of wild seed: (a) restriction of supplies to natural spawning areas and spawning seasons; (b) unpredictable supplies due to variations in the success of natural spawning from year to year; (c) the possibility of over-exploiting the wild population to the detriment of both aquaculture and capture fisheries; (d) no opportunity for selective breeding; (e) contamination with other species, some of which may be competitors, predators, and pathogens; (f) inadequate supplies acting as constraint to aquaculture expansion. The first two difficulties can be partially solved by a combination of transport and 'banking'. The latter involves storage with limited growth and is achieved for oysters by keeping them high in the intertidal zone or subjecting them to long periods of air-exposure. With young milkfish, keeping them at high densities with restricted feeding is the normal procedure. The possibility of over-fishing is sometimes recognized by the imposition of regulations on the quantities of fry that can be caught. Contamination with undesirable organisms can normally only be overcome by strategic timing of collection or by manual sorting.

In most species of cultured aquatic organisms, especially marine ones, the egg and juvenile stages of the life-history are separated by one or more *larval stages* of small size which are not cared for by the parents. The difficulties encountered in the control of reproduction may be caused by the problems of maintaining and feeding such delicate organisms under artificial conditions, or they may be caused by the inability of the adults to spawn in captivity. Whatever is the difficult stage of the life-cycle will determine the point at which the cycle has to be broken in order to replenish the cultured stock from wild populations. Since there is usually very high mortality in the wild during the early life-history stages, there is at least a theoretical advantage in collecting the young as early as it is possible to rear and protect them in captivity. This can be seen in the case of yellowtail and milkfish. In both species the fry are caught as transparent post-larvae about 15 mm long and are then reared through to fingerlings in nurseries before selling to the growers. At the time of capture they are probably about two weeks old. Young grey mullet are normally caught for culture as post-metamorphosed juveniles, although the whole life-history for this, and some other species, has been successfully replicated in captivity on an experimental scale. When bivalve culture is dependent on a supply of young stages from the wild they are usually transported and transferred to the growing facilities as *spat*, that is post-metamorphic juveniles which have settled onto a substrate ('cultch').

Irrespective of the life-history stage, the capture may be active or passive. Active netting is usual with fish and prawns (Fig. 3–1), but

Fig. 3–1   One of the methods for catching milkfish fry in the Philippines. (From S.E.A.F.D.E.C. Aquaculture Department, Philippines.)

sometimes bag-nets (fixed, conical structures) are used in tidal waters or in rivers. Some types of aquaculture rely on the young animals moving into ponds as they are flooded, for example the movement of eels, mullet and bass into the Italian coastal ponds or valli. Some mussel spat is obtained by active methods, for example by dredging in the Dutch Waddenzee. Normally, however, the collection of bivalves is passive and involves placing, at strategic times and places, special collectors onto which the larval stages settle. Since this is active habitat selection on the part of the larvae the surface nature of the collector is quite important. Sometimes natural materials are used (e.g. shell, shredded coconut husks, sticks), but plastic and other synthetic materials are often just as suitable. Thus polyethylene film has been used to capture young scallops. A fibrous substrate seems to be preferred by mussels, and ropes are very often used commercially. It seems to help if the collectors are put in the sea early and allowed to develop a hairy epifauna of hydroids and bryozoa. Once the spat has become attached to the collectors, the latter can be relatively easily transported. With mobile fish, transport can be more of a problem, but the use of plastic bags half-filled with water and then inflated with oxygen before sealing, is the technique now widely adopted, sometimes in conjunction with anaesthetics.

3.2 Control of spawning

By obtaining gametes or fertilized eggs from captive adults, a much greater control over the spawning process is achieved. One obvious advantage is that the parents of any subsequent offspring can be individually identified, which is the basis of selective breeding. Further, increased rates of fertilization and survival of the early stages can be anticipated, and it may also be possible to influence the time of the spawning process.

The degree of control will vary depending on the extent to which maturation of the gonads, mating (where appropriate), and gamete release will take place freely in captivity, and the extent to which these processes can be induced by an appropriate manipulation. If maturation is difficult, then reliance may have to be placed on obtaining ripe fish from the wild populations.

3.2.1 Spontaneous maturation and spawning

Certain species will, if grown to a suitable size, mature and spawn readily in captivity. Examples are common carp, channel catfish, tilapia, turbot, caridean prawns, crayfish and some oysters. These will all spawn in tanks, although most would normally be kept in ponds closely resembling natural habitats. In carp culture, special spawning ponds are often used. These are of various types but all provide a suitable substrate which facilitates attachment of the sticky eggs by the fish, and their subsequent collection by the farmers. After spawning, the parents are removed.

Natural spawning does carry the general disadvantage that the eggs may be difficult to collect. For this reason, *hand stripping* of the eggs and sperm from the spawners followed by *artificial fertilization* of the gametes in a bowl is often practised, especially with salmonids. The artificial fertilization of trout was first described in 1765, but was not widely practised in Europe and North America until the mid-nineteenth century.

Spontaneous spawning occurs because the species are perhaps tolerant to a wide range of factors at spawning or because key factors are provided, often inadvertently. For spawners like carp all that seems to be needed is an appropriate water temperature and substrate. In channel catfish farming, suitable containers, such as cans, drums, or boxes are placed in the ponds in April. The adults spawn in these, but they will also readily spawn in small cages or pens.

3.2.2   Environmental manipulation

In theory, it should be possible to provide all species of aquatic organisms with the right set of environmental factors for reproduction in captivity to take place. In practice this is not so because knowledge of the appropriate factors is often lacking.

However, some success has been achieved by manipulating temperature and photoperiod. Such techniques may only induce maturation, but in some cases will result in spawning as well. Besides being an aid to reproduction control in 'difficult' species, these techniques can also be used to induce spawning in spontaneous spawners outside their natural spawning period.

Grey mullet in Hawaii will develop ripe gonads if well fed in outdoor ponds, but it is more reliable to keep them in laboratory tanks and subject them to 22°C and a 6 h day-length. This procedure stimulates vitellogenesis (yolk formation in the oocytes) after 49–62 days and works independently of any preconditioning photoperiod or adjustment. In the U.K., recent application of photoperiod manipulation on rainbow trout has resulted in a 3 month advancement of spawning and similar work has also proceeded with turbot and sole. With sole the procedure is to subject the fish to 2 months of 6–8 h day-length, followed by 2 months at 16 h day-length. This results in maturation and spawning. After spawning and a 2 month refractory period, the treatment can begin again, giving two possible spawnings each year for the same fish. Temperature appears to have little direct influence on the reproductive cycle of sole, but needs to be at least 12°C in order to maintain the feeding response. Environmental manipulation can be used to induce spawning in the walking catfish by the use of artificial flooding of ponds, and in the Pacific oyster by implementing a 10°C temperature rise after conditioning in algal-enriched sea-water.

3.2.3   Endocrine manipulation

The use of photoperiod manipulation in teleosts operates on the gonads through the endocrine system. It should, therefore, also be possible to influence spawning by more direct manipulation of the hormones and considerable success has been achieved with such methods. The usual technique is to inject pituitary hormones into fish whose gonads have already matured. This is called *hypophysation*. A crude pituitary extract can be used consisting of a mixture of growth hormones, thyrotropic hormones, and sex steroids, as well as the gonadotropins which are probably the main active component.

There seems to be considerable variation in response to injected hormones between species and sexes. Often males will respond to mammalian hormones, or do not need injections at all, whereas females typically require a piscine hormone. Several species, including milkfish and grey mullet, which are not related to salmon, will respond to SG–G100 (a purified Pacific salmon extract), but carps need a carp pituitary extract. The efficiency of any dose in hypophysation is related to the sexual maturity of both donor and recipient, their phylogenetic relationship, the gonadotropin potency of the injected preparation, and the physiological state of the recipient. There is also evidence showing that the time of day at which the dose is given is important.

Either the injected hormones are regarded as exogenous gonado-tropins, or they stimulate the release of endogenous gonadotropins from the pituitary gland, or they stimulate directly the natural steroids or other compounds in the gonad (Fig. 3–2). Steroids themselves can also be

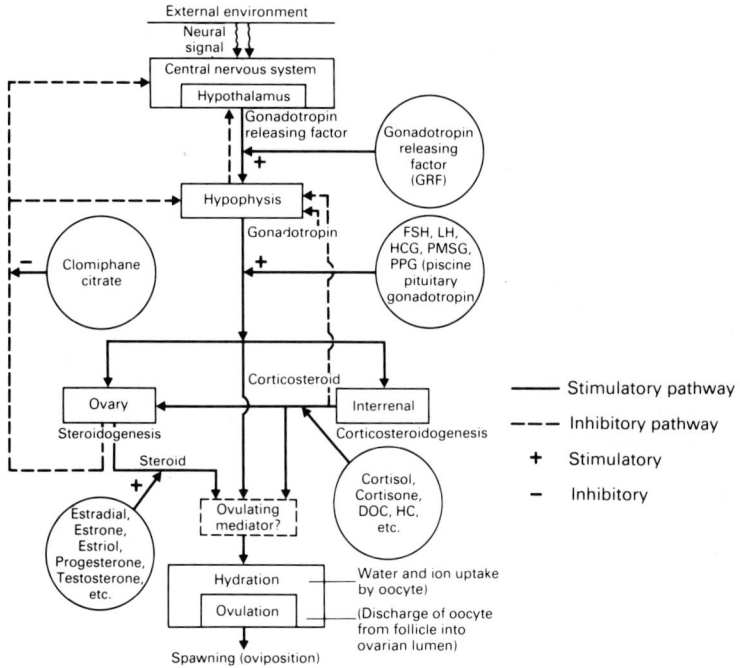

Fig. 3–2 The interrelationship of endocrine organs influencing ovarian development and ovulation in teleosts. The effective hormones are circled. (After KUO, C-M. and NASH, C. E. (1975). *Aquaculture*, 5, 19–29.)

injected. Some of the best-documented work on induced spawning has involved grey mullet at the Oceanic Institute, Hawaii (KUO *et al.*, 1974).

There will undoubtedly be many advances in the use of hormones to induce spawning in fish in future, both in terms of techniques and species. One technique may be to use slow-release capsules implanted within the fish, as an alternative to the potentially stressful injections. Only in 1977 (in the Philippines) was the first induced spawning and artificial fertilization of milkfish achieved by hormone injection; a highly significant result in view of the economic importance of this species. Similar success was achieved with eel, *Anguilla japonica*, in Japan in 1974, but both results are a long way from the large-scale application now achieved for the Indian and Chinese carps (CHAUDHURI, 1976; ANON, 1977) and perhaps soon to be developed for grey mullet.

Penaeid prawns are typically reluctant to mature and spawn in captivity, the main exceptions being *Penaeus merguiensis* and *P. japonicus*. Induced spawning in at least some other species can be achieved by removing the eyestalk in the middle of the intermoult period. This eyestalk ablation is a form of endocrine manipulation because the eyestalk contains the production and storage sites – the X-organ, and the sinus gland respectively – for the gonad-inhibiting and moult-inhibiting hormones. The most recent, and commercially most significant success has been with the largest and fastest-growing penaeid prawn species, the jumbo tiger prawn.

3.2.4   *Inhibition of spawning*

Normally in aquaculture, seedling production and growing-on (to marketable size) are kept as separate activities. If the growing stock begin to spawn, therefore, this is considered undesirable. Sometimes spawning actually takes place and leads to overpopulation of ponds and consequent stunting of growth in tilapia species. Alternatively, the diversion of energy into gonad development will itself have an adverse effect on growth-rate of the stock and perhaps also on the condition of the carcass. This is a problem in farming Atlantic salmon since a proportion of the males have to be harvested at small size due to early maturation of gonads. That is, they become *grilse*. There seems no certain remedy at the present time, although work on genetic, immunological, and environmental control is proceeding.

Inhibition or prevention of tilapia breeding/overpopulation can be achieved in the following ways: (a) *single sex stocking* (usually males, as these have the added advantage of a faster growth-rate); this is done by either manual sorting, incorporation of methyltestosterone in the diet, which converts males to females, or an appropriate interspecific hybridization which gives up to 100% male progeny; (b) *cage culture*; most tilapias build a nest on the pond bottom, and if prevented access to the bottom, will either not breed, or the eggs will be lost through the cage meshes; (c) *co-stocking with a predator* (biological control).

3.3   Hatcheries

Once fertilized eggs have been obtained, the next step is to incubate them to hatching and then rear the larvae to beyond metamorphosis when transfer to the on-growing facilities usually takes place. The hatching stage will be by-passed in those species which brood their own eggs, and in situations where the young are obtained from the wild, although the latter do require special care before transfer to on-growing. Commercial hatchery production of many freshwater species, in particular cyprinids and salmonids, is widespread, and details can be found in BARDACH *et al.* (1972). Only two points will be emphasized here in relation to salmonids: first, the use of artificial pelleted diets from the time of first feeding (*swim-up stage*), and secondly, the convenience and success of transporting these species at the *eyed-egg stage*.

Hatchery operators are concerned with the growth and survival of the young organisms, so there are many parallels here with the on-growing stage described in Chapters 4 and 5. Nevertheless, the youngest stages of all species are more sensitive and difficult to maintain than post-metamorphic stages and have requirements that can best be satisfied by the application of the specialized techniques available in the hatchery. This particularly applies to those marine species with a small planktonic larval stage in the life-history. High *natural* mortality in the early stages of such species inspired the early hatchery operators of the nineteenth century to attempt to increase overall survival by protecting larvae in the controlled, predator-free conditions they could provide. However, in contrast to the case of salmonids, the problem of feeding the young fish and thus encouraging growth to a suitable size was such that they had to be released back into the natural environment soon after resorption of the yolk sac. Although it is probable that no net benefit to the stocks resulted, the experience did provide the foundation for current hatchery practice.

3.3.1   Marine teleost hatcheries

Marine teleosts are generally more difficult to rear than salmonids and an instructive example here is to consider the U.K. research and development over the last 25 years on the hatchery production of young flat-fish (Pleuronectiformes). The work was started at Lowestoft, and SHELBOURNE's (1975) review can be consulted for further details.

The initial work was carried out on plaice using an elaborate closed-circulation system. The first fish to be raised beyond metamorphosis were obtained in 1957, and by 1961, 5% survival and a final density of 336 fish m^{-2} had been obtained. The cultivation unit then moved on to Port Erin, Isle of Man, where eggs could be obtained from a captive breeding stock, and where the seawater quality was good enough for an open seawater circulation to be used. With these improvements and the use of antibiotics, such as penicillin and streptomycin, to treat the eggs, 66% survival to final densities of 7570 m^{-2} were possible.

The post-larvae were fed on live nauplii of the brine-shrimp, *Artemia salina*, an organism which is now of great importance in hatcheries throughout the world. Two other techniques improved the quality of the young fish and hence reduced mortality: (1) only acclimatized, unstressed parents were selected, since the quality of the spawners was reflected in the quality of the progeny; (2) size segregation of the growing young helped to prevent growth inhibition of the smallest fish, and also fin-biting which could lead to disease and osmoregulatory difficulties.

Direct commercial application of the plaice hatchery work, however, did not develop. It was shown by JONES (1972) that the farming of this species would not be economic, although two other flat-fish species, sole and turbot, would be promising candidates. Attention is now focused on these species, and this includes a consideration of hatchery techniques and problems. A hatchery survival rate of 10–20% appears to be economically adequate (KERR, 1976) and this can now be achieved for turbot and sole, at least on a moderate scale.

The problems have been mainly associated with feeding. Turbot hatch at 3 mm body length in contrast to the 5–6 mm of plaice. Whereas the latter were able to take *Artemia* nauplii (0.2 mm) at the beginning of exogenous feeding, turbot required a smaller animal for the first few days. The one chosen was a rotifer, *Brachionus plicatilis* (0.1 mm). Current difficulties, centre around the early post-metamorphic stages, and the transition from natural to artificial diets.

More is probably known about rearing young flat-fish than other groups of marine teleosts, though experimental work is proceeding on a variety of species throughout the world. In Japan, there is large-scale hatchery production for red sea bream (*Sparus major*) mainly used for stocking the Seto Inland Sea. A useful review of problems of marine teleost hatcheries, with particular reference to grey mullet is given by NASH and KUO (1975).

3.3.2 Shellfish hatcheries

Hatcheries for the mass production of post-larvae of the giant freshwater prawn have been established in several countries. It was in Malaysia that the techniques were first developed. In 1961 at Penang, it was found that the developing larvae required a salinity of 12–16%, rather than fresh water, to complete their development. This prawn belongs to the caridean group, whose newly hatched larvae feed on live animal food (such as *Artemia* and *Brachionus*) and prepared diets, and have no requirement for algae (WICKINS, 1976). They do, however, have a tendency towards cannibalism when kept at the high densities necessitated by hatchery procedure, though it is more of a problem after metamorphosis.

Penaeid prawns require an algal diet for their protozoea larvae, which precede the carnivorous mysis and post-larval stages. Penaeids are generally more fecund than carideans and are less liable to be cannibalistic, but hatchery successes are few. Only those for *Penaeus japonicus* are at present capable of mass production and they provide seed for stocking inshore waters in Japan and also for intensive culture.

Lobster hatcheries operated in the nineteenth century, but, largely because of high mortalities caused by cannibalism, the larvae were released into the sea at an early stage. Work on both American and European species proceeds at an experimental level. A related species, the freshwater signal crayfish is currently the subject of commercial production at a hatchery in Sweden which exports juveniles to many countries in Europe.

Oyster hatchery techniques are well established (LOOSANOFF and DAVIS, 1963). One system for the Pacific oyster, described by O'SULLIVAN and WILSON (1976), involves rearing the larvae in polythene bins, using filtered ultraviolet sterilized seawater at 23–25°C. Embryos are initially placed in the bins at 100 ml^{-1}, but after hatching into veliger larvae, they are sieved, sorted, and kept at 10 ml^{-1} while fed with a mixture of algae including *Chaetoceros calcitrans* at 30–100 cells μl^{-1} day^{-1}. Eventually they are screened into different size-groups and maintained at 3–5 ml^{-1}, and from 3–4 weeks after fertilization, larvae larger than 210 mm are removed to the shell-settlement unit. Here the larvae metamorphose and settle onto autoclaved shell particles (200–300 μm in size) in trays, and after 10 days the trays are screened and the spat collected on 500 μm and 350 μm sieves, which are then transferred to the nursery. *Tetraselmis suecica* becomes the preferred food, and the spat are kept in the nursery trays until about 5 mm in diameter. This method produces unattached or cultch-free young oysters.

3.3.3   Culture of live foods

Unicellular algae are widely cultured as live foods, especially for oyster larvae. Even if they are not fed upon directly by cultured larvae, their presence in hatchery containers is thought to benefit the rearing environment for several species through the removal of metabolites or the supplementation of necessary vitamins or amino acids in solution.

In addition to algae, the most convenient and widely used live food is the brine-shrimp, *Artemia salina*. This food is also valuable to aquarists, since the dried eggs or cysts, harvested from salt lakes in America, are easy to hatch out into the actively swimming nauplii. In aquaculture, large-scale production of nauplii is required, and several automatic systems for mass culture have been developed, producing for example, 250 million per day at a maximum loading of 1 g litre^{-1}. The increasing cost of the cysts, and unreliability of supplies have led to a closer study of the animal and its use in order to prevent its becoming a bottle-neck to aquaculture development. Most of the current work is being carried out at the University of Ghent in Belgium (e.g. SORGELOOS and PERSONNE, 1975). It has been found that there are many different strains which vary in environmental requirements and suitability as a larval food. Further, different batches of the same strain vary in quality due to the different histories of the cysts in the salt ponds (mostly in California). The food value of unfed nauplii is maximal immediately after hatching and declines by up to 28% during the first 24 hours. Their suitability as food is also diminished if the cysts are not separated from the hatched nauplii since these can cause intestinal blockages (and death) if ingested by delicate fish or prawn larvae. Separation can be achieved by exploiting the phototactic swimming behaviour of the newly hatched nauplii, or by decapsulation of the unhatched cysts using sodium hypochlorite.

3.4   Improving the quality of seed through genetics

Compared to agricultural livestock, little is known about the genetics of fish and shellfish. What is known has only rarely been applied commercially, since most attention has so far been given to adapting the environment to suit the organism rather than the other way around. One major limitation to genetic improvement is that unless the organism will complete its life-cycle under controlled conditions, no real progress can be made. Only common carp and rainbow trout are considered to have domestic varieties at present. The carp has been cultured for thousands of years, and the most distinct varieties to emerge are the coloured ornamental or koi carp from Japan. But some of the carp cultured for food also differ morphologically from their wild Danubian ancestors and are familiar as mirror or leather carp with their pattern of reduced scaling. The important characteristics of the domestic carp, are its faster growth, deeper body, larger mouth, longer intestine, and better utilization of plant food. There are also distinct varieties in S.E. Asia, perhaps developed from a separate sub-species and sometimes known as Chinese big belly.

Genetic research and selective breeding of the carp continues today, particularly in Israel and the U.S.S.R., and it probably still receives more attention than any other aquatic species. Objectives of the work include improving conversion efficiency, taste, the edible percentage of the carcass, and resistance to unfavourable environments and disease. Some of the results are discussed in sections 4.5 and 5.5. Here the available techniques are briefly considered.

1   *Selection.*   The genetic basis for selection in fish is poorly understood, and according to MOAV (1976) 'it cannot be over-stressed enough that phenotypic or even overall genetic variation is not sufficient for a successful selection programme'. In order to benefit from simple selection techniques, genetic variation needs to be of the additive type. This may apply, for example, to egg size and age at maturity, but the available evidence suggests that it does not apply to growth. It has been suggested that the most effective method of selection may be to combine family with mass selection and use males selected by their progeny.

2   *Hybridization.*   Fish seem to hybridize readily. This offers the possibility of improved growth through F_1 hybrid vigour; introgression; the creation of a wider gene pool as a basis for further selection; combination in a single fish of qualities of different genotypes; and the possibility of reproductive inhibition through sterile and monosex

progeny. Hybrids may be intergeneric, interspecific, or intraspecific. One of the best known in the former category is the 'bester', a hybrid between the sterlet (*Acipenser stellata*) and the much larger beluga (*Huso huso*). The hybrid is produced by artificial fertilization from stripped gametes and has a wide range of salinity tolerance.

An example of beneficial intraspecific hybridization concerns the Chinese and European races of the common carp. Under conditions of polyculture and cheap food and manure inputs, the hybrids combine the European characters of fast growth and late maturity with the Chinese characters of resistance to crowding and ability to utilize low-grade food. There is also some heterosis or hybrid vigour.

3   *Gynogenesis and Polyploidy.*   Many aquatic animals are more akin to plants than terrestrial livestock because of their high fecundity and the fact that both male and female gametes can be handled outside the organism, thus facilitating manipulative control over the developing zygote. In gynogenesis the sperm head has to penetrate the ovum to induce embryo development from an exclusively maternal nucleus, but the sperm is inactivated by irradiation. Diploidization of the female chromosome set is achieved by cold shock. The advantage is that a single superior genotype, regardless of its level of heterozygocity can be maintained and multiplied. Work has been carried out at the MAFF Lowestoft laboratory on gynogenesis in flat-fish (PURDOM, 1972) and although it has no commercial application at this stage, the technique is effective and is of potential value for quickly producing inbred lines. Another opportunity, but one which does not commonly occur in fish, is afforded by the threadfin, *Polydactylus sexfilis*. This is a protandrous hermaphrodite, and so by cryopreserving sperm and then self-fertilizing, the gynogenesis effect can be achieved. Other relevant hermaphrodites include flat oysters and scallops.

Cold-shocking diploid eggs and zygotes may produce autopolyploid offspring that should be sterile, and may grow faster than diploids. Triploid plaice were produced in this way at Lowestoft, and both sexes were sterile. Work on salmon has shown that tetraploid fish can be developed by treating the newly fertilized eggs with cytoclasin B. Hybrids between tetraploids and diploids could give sterile triploids, which in the context of early maturation (§ 3.2) could be of commercial importance.

4 Maximizing Growth

4.1 Aspects of fish growth

Once young fish of a suitable size have been obtained, they are ready for stocking into the on-growing facilities. During the on-growing phase the farmer is interested in maximizing his yield from the initial seedstock and this is achieved by maximizing (a) growth and (b) survival. The latter will be considered in the next chapter, but it should be appreciated that measures taken to maximize growth will usually result in good survival. Growth is typically measured as the gross body weight but may be refined to refer to weight of edible parts, weight of protein, dry weight or units of energy. There are two relevant, but distinct, aspects to growth: (1) growth rate; (2) growth efficiency (i.e. food conversion efficiency).

Growth rates vary markedly between and within species, and it is important to emphasize the plasticity of growth in response to the environment in the poikilothermic aquatic organisms under consideration. Temperature is by far the most important abiotic factor affecting growth, and this is referred to again below and in section 4.5.

Perhaps the fastest growing fish under culture is the grass carp which can grow from 20 g to 2.5 kg in only 6 months and to 8.5 kg in a year. In general, farmed tropical organisms reach marketable size in less than a year and so more than one crop can often be grown within each annual cycle. Thus milkfish in S.E. Asia, marketed at 200–300 g, can give 2–3 crops a year. In temperate waters, by contrast, each growth cycle lasts more than 1 year and sometimes 2 to 3 years.

Individual differences occur even amongst siblings and this can be a problem because in many species size hierarchies develop, with the largest fish inhibiting the growth of the smaller ones and sometimes eating them. For this reason it is common practice to periodically grade the stock and separate the different size categories, although this does not prevent individual growth differences from occurring. The relative size of common carp immediately after hatching is a major determination of further growth rate, but the cause of the initial advantage (genetic or environmental) is poorly understood. Any difference in growth rate between two individuals or two species must, however, be due to one either eating more food than the other or utilizing ingested food more efficiently (or both).

The latter is normally measured by *gross growth efficiency*. Expressed as a percentage, this is given by (growth/intake) x 100. Values of 15–28% are common in finfish culture and are generally higher than those for terrestrial livestock. The highest values recorded in fish are 77–79% but this refers to yolk utilization in the early larval stage of the sardine (*Sardinops caerulea*). Nevertheless, it does indicate a maximum value against which attempts to improve conversion efficiencies can be measured. Instead of gross growth efficiencies, the fish farmer usually uses *feed conversion ratios*. This is given by food given/growth, and so a 30% efficiency would be equivalent to a 3.3 : 1 ratio. It assumes that the food given to the fish is all eaten (and represents their only food intake) and that the same units are used in each case. Information on the first point is usually lacking, but with regard to the second, the ratio is often in terms of dry weight of food and wet weight of fish which is why ratios of 1 : 1 are frequently reported.

When feed costs are an important component of production costs, as they are in most intensive aquaculture, then good conversion efficiency becomes an important criterion. Given a choice, it may be better to have a slow grower which utilizes its food efficiently than a fast grower which is very wasteful of food. However, the sooner the stock reaches marketable size the less the risk of stock loss, the quicker the return on initial investment, and the greater the use of holding facilities.

Because of the economic significance of conversion efficiency, it is important to understand more precisely the relationship between food intake and body growth. This is best done by considering either energy or nitrogen budgets. The following is the usual form of the energy budget:

$$C = F + R + U + P + G$$

C	F	R	U	P	G
consumption, or food intake	faeces	metabolism	excreta	body growth	gonad growth

or

$$P = C - (F + R + U + G)$$

Total metabolism can be broken down into standard (basal), active and digestive components. The latter, sometimes called specific dynamic action, occurs whenever food is digested and assimilated and is highest for proteins and for large infrequent meals, so its magnitude is open to manipulation.

Poor conversion efficiency then may be due to low assimilation, a high rate of metabolism or to the partitioning of food materials into gonads and perhaps other products. The object of growth maximization must be to ensure a high food intake, and the conversion of as much as possible of that food into edible fish flesh (muscle) and it can be seen that the various stages in the conversion of food to flesh are open to manipulation through feed, environmental factors and the fish themselves. (Table 4.)

One of the most valuable recent studies of a teleost energy budget is that for the brown trout by ELLIOTT (1976). Some of his results are summarized in Fig. 4–1, but consultation of the original papers is recommended in order to gain insight into both food intake and

Table 4 The likely processes through which various factors directly influence intake, assimilation, metabolism and gonad growth, and hence growth.

	Temperature	Salinity	pH	Dissolved oxygen	Water flow	Light	'Pollutants'	Parasites	Available food	Food quality*	Age/size of fish	Density of fish	Sex of fish
Intake/consumption (C)	x	x	x	x	x	x	x	x	x	x	x	x	x
Assimilation (C–F)	x			x					x	x	x		
Standard metabolism (R_s)	x	x	x				x	x			x	x	
Active metabolism (R_a)	x	x	x	x	x	x	x	x	x	x	x	x	x
Specific dynamic action (R_d)	x								x	x			
Gonad growth (G)	x					x			x	x	x	x	x

* includes chemical composition.

partitioning in a fish and into the techniques and problems of obtaining the relevant data. The endocrine control of growth in fish is receiving increasing attention. Anterior pituitary growth hormone and thyroxine are known to be involved, but other hormones, such as insulin and prolactin, are important because they affect the release and storage of lipids. There is clearly considerable scope for improving growth by endocrine manipulation.

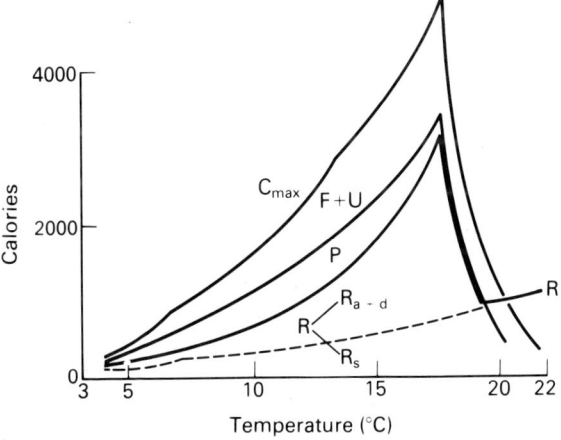

Fig. 4–1 Effect of temperature on the major components (R_s, standard; R_a, active; R_d, digestive metabolism) of the energy budget for brown trout (50 g live weight) on maximum rations. (After ELLIOTT 1976.) (1000 calories = 4.2 MJ.)

Further information on fish growth and its importance in aquaculture is provided by WEATHERLEY (1976).

4.2 Nutritional requirements

Once nutritional requirements have been defined it is then possible to optimize growth by manipulating the chemical composition of the diet. However, there are considerable difficulties in defining these precisely, partly because they vary with the size and condition of the fish and with the environment in which it is living, and a major problem to date has been a lack of standardization. This has been succinctly exposed by NEW (1976) in his proposal for the standardization of experimental conditions for dietary trials of prawns. Here attention was focused on animal origin, trial length, measurement technique and frequency, pre-trial maintenance, trial environment, replication of experiments and food preparation and feeding technique.

Tentative nutritional requirements have been determined for only a few species, such as rainbow trout and channel catfish. These are with respect to the ten *essential amino acids, gross protein, water and fat soluble vitamins* and some *essential polyunsaturated fatty acids* of the omega-3 series. The estimated requirements, plus a 50–100% safety factor, have been incorporated into diet ingredients for least cost computer formulation of practical artificial feeds (§ 4.4). If real progress is to be made with diets, more basic nutritional studies need to be carried out not only on the values of specific nutrient requirements, but also on digestibility co-efficients, the sparing effect of one nutrient upon another, and the role of gut bacteria.

Protein is the most significant component because of its high cost and because it is an important constraint to growth. The gross digestible protein requirements for a number of species have been determined and mostly these range from 28–50% of the total diet weight. Much research is being undertaken to reduce the protein content without adversely affecting food conversion efficiency. This means increasing the protein efficiency ratio and ensuring that the protein supplied in the diet is used for growth and spared as an energy source. As well as being wasteful, protein levels in excess of requirements may actually bring about a depression in growth because of the high energy cost (specific dynamic action) of their digestion and because of the excretion of large quantities of ammonia which may cause stress and severe gill damage in confined situations.

An alternative to reducing the protein requirement is to utilize cheaper sources of protein, (§ 4.4), but here the quality of the protein – the amino-acid composition – becomes a critical consideration.

Nutrition is one area of aquaculture where much more biological information is required. Reviews for prawns (NEW, 1976) and teleosts (HALVER, 1976) can be consulted for further details.

4.3   Natural feeds and the role of fertilizers

Even in intensive aquaculture, some natural food is sometimes available; for example, in sea-cage culture of salmon the fish can make some use of the larger planktonic organisms drifting through the cages and the small fish and crustaceans actively attracted to the site by food and shelter. Although the natural food is inadequate to support the high stock densities involved in such situations, there is no information on the actual contribution made and to what extent it improves the conversion ratios recorded for the artificial feed. However, as indicated in Table 1, some types of animal aquaculture can be wholly dependent on the *in situ* natural production, and no artificial feeding is necessary. These are: (*a*) culture of filter-feeding bivalves; (*b*) certain types of pen (milkfish) and cage (tilapia) culture of fish in freshwater; (*c*) certain types of non-intensive pond culture of fish and prawns; (*d*) stocking or ranching.

In (*a*) and (*b*) and often in (*c*) the stock are feeding on plants and it is this which enables such high stocking densities and production to be maintained without a large feeding subsidy. The milkfish are feeding mainly on the phytoplankton which is of course also the food of bivalves in culture. The caged tilapia feed on phytoplankton, zooplankton and the algae growing profusely on the cages. Fish in ponds may feed either on the plankton or on the benthic and floating algae and macrophytes – and also on benthic and planktonic invertebrates. In situations where natural food is important, the presence of large numbers of competitors will reduce the growth of the species under culture. These are usually other species of fish in ponds or sessile invertebrates in bivalve culture. However, in milkfish ponds, snails, polychaetes and chironomids are also important competitors with milkfish for the algae and are normally controlled chemically.

Although the systems so far described do often operate within the constraints imposed by natural production they are open to further manipulation either indirectly, by improving primary production or, directly, by supplementary feeding. (The latter will be considered in the next section.) Sometimes benefits are possible through attention to the substrate, although this will not so much increase overall primary production as influence the dominant plant types. Thus in brackish water milkfish ponds either phytoplankton, filamentous algae, or the benthic algal felt known as *lab-lab* may develop. The latter is the preferred food for the milkfish and its development is critically dependent on water depth and water clarity. In order to encourage lab-lab, there is a sequence of draining and filling the ponds from November to February before the first fish are added.

The main method of stimulating primary production in ponds is through nutrient addition using either inorganic or organic fertilizers. *Inorganic fertilizers* contain concentrated amounts of at least one of the three major plant nutrients; nitrogen, phosphorous and potassium. It is generally acknowledged that phosphorous is the element most likely to limit production in fish ponds, and that potassium is only likely to be important in very peaty soils. More doubt is associated with the role of nitrogen, although nitrogenous fertilizers are commonly a component of fish-pond fertilizers. An important factor here may be the role of nitrogen-fixing blue-green algae which occur in the plankton but are most significant as a component of lab-lab flourishing in the soft, hydrophilic, biologically active, shallow pond muds.

Although not strictly a fertilizer, it is convenient to consider liming here. Lime ($CaCO_3$) is commonly added to freshwater ponds and is considered necessary if base saturation of pond mud is less than 80% and total hardness of the water is less than 20 mg l^{-1}. If hardness is less than 10 mg l^{-1}, the addition of fertilizers will not be effective in stimulating plant growth, since lime promotes the release of nutrients and the breakdown of organic matter. Lime also buffers and stabilizes pH and reduces the incidence of certain diseases. When applying inorganic fertilizers to stimulate production two points need to be considered: (1) overstimulation may occur resulting in phytoplankton blooms with associated toxicity, oxygen and shading problems; (2) nutrients applied may be adsorbed onto the mud before they can be taken up by the plants; liming is important here, as is the periodic draining and drying of ponds to oxidize accumulated mud and release the nutrients.

Some interesting experiments were carried out in Scottish sea lochs in the 1940s to determine whether marine fish production could be enhanced by fertilization with superphosphate and sodium nitrate. There was some suggestion of an improvement in flatfish growth through an increase in production of benthic food organisms, but the evidence was not conclusive.

Animal and human dung and various plant wastes are the traditional *organic fertilizers* for fish ponds, and there is much current interest in systems which are able to dispose of waste by converting it to valuable protein. Details are given by PASTAKIA (1978). In practice, organic and inorganic fertilizers are often both applied to the same pond. Weight for weight organic fertilizers contain less of the major plant nutrients, may be more costly to transport and apply, and can give rise to oxygen-depletion problems. However they are likely to contain a wide range of nutrients, and improve pond-soil structure and are a source of food as well as fertilization.

The inorganic nutrients which are released by decomposition of the organic matter become available to phytoplankton and other plants. The decomposers (such as bacteria and protozoa) themselves also form the start of a food web which is not dependent for its inception on sunlight, a fact which may be of some relevance when the water is turbid and the weather cloudy. Studies have shown that crops of zooplankton and chironomid larvae (important fish foods) can be ten times higher in manured than in non-manured ponds, and it has also been demonstrated that *Daphnia* spp will ingest manure particles and digest the bacteria on their surface.

Ignoring the public health and acceptability problems associated with the use of human sewage (in China the former is now reduced by anaerobic treatment of the sewage before adding it to the ponds), the main limitation to the use of organic wastes is imposed by the biological oxygen demand (BOD) generated by the aerobic decomposer populations. In Israel, however, it has been found that rates as high as 1.5 t ha^{-1} day^{-1} of liquid cow manure can safely be added. Very high rates of fish production (up to 32 kg ha^{-1} day^{-1}) have been recorded from static Israeli ponds with cow manure replacing pelleted food, and it is estimated that 10 t of dry manure are converted to 4 t live-weight of fish. Frequent spreading of the manure as a slurry over the pond surface enables much larger quantities to be applied than in the old method of heaping or spreading in one go on the bottom of the drained pond (limited to 2 t ha^{-1}).

Ducks can be regarded as living manure-spreaders. Five hundred can be kept on a hectare and each produces 6 kg droppings per month. One hundred kg fresh duck manure produces 4–6 kg fish flesh. Some fish farms involve the keeping of animals, particularly pigs, adjacent to or over the ponds so that dung, and any uneaten food fall straight into the water. Human dung, as night-soil, often reaches the ponds in this way from strategically placed latrines, a common method of sewage disposal/protein production in parts of S.E. Asia. Some current experiments use treated or partially treated effluent rather than raw sewage and sophisticated tank systems rather than earth ponds, but the principles are the same. An example involving polyculture is shown in Fig. 4–2.

4.4   Artificial feeding

In artificial feeding, materials are provided for direct consumption by the stock either as a supplement to or as a complete replacement for a natural diet. The materials range from additional supplies of natural foods (such as aquatic plants for the grass carp) to formulated diets (such as the pellets fed to salmonids).

Whatever is added to the pond should be cheap (in relation to the market value of the cultured species), acceptable to the stock, ingestible and digestible, and of a chemical composition that will encourage fast growth, good conversion and survival. Organisms forming part of the natural diet are often used in the initial stages of aquaculture development; for example, the use of sandeels (Ammodytidae) and anchovies (Engraulidae) and other trash fish in Japanese yellowtail culture. A slight departure from a strictly natural diet is the use of marine trash fish for freshwater walking catfish and bullfrog (*Rana catesbiana*) farming in Thailand. Added food organisms are either caught or cultured specifically as a feed, or are a by-product of some other industry. To some

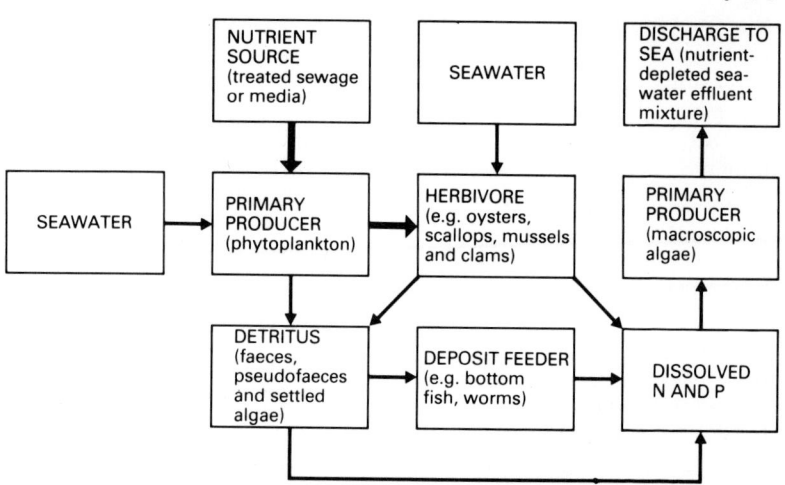

Fig. 4–2 Model of a multi-species aquaculture ecosystem using treated sewage as a source of nutrients. (After HUGUENIN, J. E. (1975). *Aquaculture*, **5**, 135–50.)

extent the latter may be regarded as wastes, (e.g. trash fish, silkworm pupae) but more of the trash fish landed in Thailand is now being bought up for conversion into fish meal. As a result the decreasing quantity and higher price place a real constraint to the expansion of catfish farming using this source of food. Raw trash fish has been used in Japanese yellowtail and eel culture and in Danish trout culture but the associated handling, water quality and disease problems make formulated diets, usually containing fishmeal, a more attractive solution for these carnivorous species.

On a global scale, probably most of the feed added to ponds is of plant origin and the increasing cost of even low-grade animal protein will in future make it even more important. Most of the plant material currently fed to fish are by-products from agriculture, are used as supplementary feeds, and consist of items that the fish would not be expected to meet in their natural environment. Rice bran, soybean meal, peanut meal, ground-nut oil cake, and wheat middlings are some common items added as supplementary feed to fish ponds in Israel, India and S.E. Asia.

PITCHER (1977) emphasized the high energy cost of a fish meal component in trout pellets and recommended consideration of cheaper protein sources, and it has been shown that trout grow perfectly well when soya bean meal replaces 40% of the fish meal in the diet as long as some additional cystine and tryptophan are incorporated. A complete replacement of fish meal by soya may be possible, if the soya is made more digestible by prior heat treatment. Common carp grow as well on a mixture of 68% of fish-meal-free basic diet and 32% of the green alga, *Scenedesmus obliquus*, as they do on commercial feeds containing fish meal.

It is only when the algal content reaches more than 50% that loss in weight and deficiency symptoms occur. Single cell protein (either derived from hydrocarbons or as a component of activated sewage sludge or distillers' dried solubles) is also being investigated as a fish-meal substitute, and so are poultry by-products and hydrolyzed feather meal. The latter can form up to 30% of the diet for trout (compared with only 5–10% for pigs and poultry) before a growth depression sets in. Successful use of poultry by-products requires supplementation with small quantities of lysine, tryptophan and methionine.

Commonly, mixtures of both plant and animal feedstuffs are used in formulated diets. A typical trout diet could include bone meal, fish meal, dried skim-milk, soya meal, wheat middlings, brewer's yeast, cellulose flour, vitamin mixture and feeding oil, and may have an overall composition of protein (45%), oil (7%), carbohydrate (23%), fibre (4%), ash (12%) and moisture (9%). Diets, in addition to providing *nutritional requirements* may carry *colourizers, drugs, hormones* and *feeding stimulants*. Thus cantaxanthin is added to salmon diets at 50–100 mg kg^{-1} in order to colour the fish flesh pink. It also seems to improve overall performance of the fish. Cobalt chloride and chloramphenicol are added at 10 mg and 1 mg kg^{-1} respectively to some carp foods to control infectious dropsy. The use of 17 -α methyltestosterone in diets has been used to enhance the growth of young Pacific salmon and bring about sex-reversal in tilapia species (§ 3.2). Binders have to be used in most artificial diets, and starch is the most commonly chosen.

Some feeds are given in their natural form, but most are transformed in some way. Various processes are used in the transformation, but the end result will reflect ease of transport, storage and application as well as acceptability to the animals. Hard dry pellets have many advantages and are almost universally used in modern salmonid culture. Although such pellets store well, some fish do not readily accept them unless they are softened before use or unless moist pellets (which can be stored frozen) are used. The Oregon moist-pellet is often used for young salmon: dry meal is mixed with 30–40% by weight of fresh or frozen fish and the mixture is forced through holes in a plate as in macaroni manufacture. Similar moist feeds are favoured for turbot and sole. Pastes are used in Japanese eel culture.

When pellets are being used their size is an important consideration. As large a pellet as the fish can take is used, since this will be cheaper and the fish will use less food-gathering energy per weight of food consumed if they eat fewer larger items. Pellets may either be floating or sinking and the choice will depend on the species and the type of holding facility. Most pellets seem to disintegrate in the clutches of feeding prawns, and some progress has been made with micro-encapsulated feeds for this group.

Feeding stations are sometimes used as an alternative to spreading food over the whole area of the holding facility, and this will save time if the latter is large. It applies to some European carp ponds and also to eel culture (Fig. 4–3). Such methods depend on the fish learning where to come in order to obtain food, and in some experiments in a Scottish loch, trout have been conditioned by an accompanying sound stimulus. Hand-feeding tends to be time-consuming but it is still preferred even on some big farms because an experienced stockman will learn a lot about the condition of the fish by observing their behaviour at feeding time. *Automatic feeders* are, however, widely used where the food is in pellet form. They may be pre-set to release a calculated amount of food at fixed intervals but *demand feeders* are increasingly used. Here the fish themselves determine the rate of release from the feeder by depressing a trigger.

Fig. 4–3   Feeding eels in Japan. Food is in the form of a paste which is placed in a basket and lowered into the pond. The eels are thereby attracted to the feeding station and swim through the meshes of the basket to feed.

In situations where the farmer rather than the fish determines the amount of food, reference is made to the age and biomass of the stock present (not always exactly known) and factors such as temperature which affect feeding rate. If an excess of food is given and it remains in the system, it will decompose and may create an oxygen problem. The presence of other species such as crustaceans, acting as scavengers but eventually forming a secondary crop, will help to make better use of any excess food. Trout are given a daily weight of pellets which is up to 5% of biomass present. Looking at the amount of food needed over the whole on-growing phase about 150 t of pellets would be needed for a farm producing 100 t fish. Eels may be fed what they can eat in 20 min each day and this amounts to 5–15% of their body weight of trash fish and 1–3.5%

of artificial paste diet. Normally fish are fed more than once a day and in trout farming the daily ration is split up into as many as eight separate feeds. This allows more efficient digestion and assimilation.

4.5 Manipulations of stock and environment

So far only diet manipulation has been considered as a way of increasing growth and production. Improvements can, however, be made by manipulating the stock to suit a particular set of environmental conditions or by manipulating the physical environment to suit the stock.

4.5.1 Stock manipulation, including genetics

This begins at the stage of species selection (§ 2.1) which is relatively straightforward where monoculture is concerned but much more complex when a species mix has to be balanced in polyculture. There is plenty of scope for improvements in polyculture stocks, but the benefits of using more than one species, particularly in ponds where natural food is important, have been convincingly established. This is shown, for example, by five species of Chinese carp (CHEN, 1976) which are stocked together to make optimum use of the pond ecosystem since they feed at different levels in the water and on the different food sources available (macrophytes, phytoplankton, zooplankton, detritus and benthic invertebrates). The different species in a polyculture system also respond differently to the addition of fertilizers and supplementary feed (Fig. 4–4) providing further impetus to careful selection of the initial stock.

Even in monoculture the initial stocking density, size and sex of the stock, their condition and previous history will affect their growth-rate, and the work of stock manipulation continues throughout the on-growing phase. A multiple-stocking programme may be chosen, and

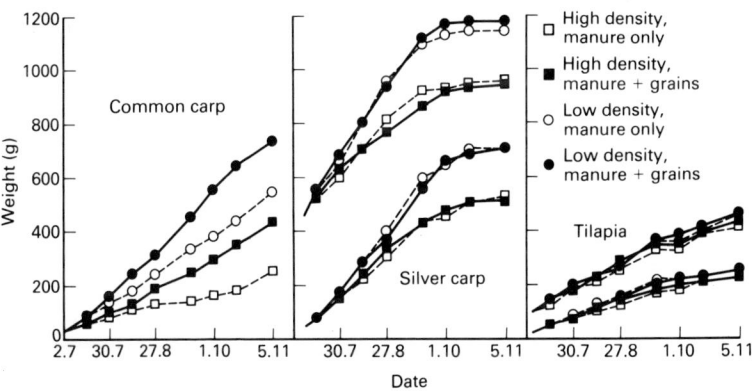

Fig. 4–4 Growth curves of common carp, two groups of silver carp, ♂ *Tilapia aurea* and *T. aurea* x *T. vulcani* hybrids in static ponds receiving 2 treatments at 2 stocking densities. (After MOAV *et al.*, 1977.)

thinning, size segregation and multiple-harvesting can be considered as further fine-tuning of the stock to the food and environmental conditions of the system.

Best use of the system will be made if stocking density is as high as possible, but this means that because of growth, either the initial stocking must be suboptimal or a high initial density must be followed by thinning and transplanting to larger ponds (Table 5). Alternatively there may be

Table 5 A typical procedure of stocking a single size-group of milkfish in ponds of different sizes. (After TANG, Y. A. (1972). In: *Coastal Aquaculture in the Indo-Pacific Region*, edited by T. V. A. Pillay. Fishing News Books, Farnham, Surrey. 438–53.)

Stage of growing	Size (ha)	Ponds required		Initial pop. size			Number of days cultured	Expected pop. at end of growing stage	
		Number	Total area (ha)	No/ha	No/kg	kg/ha		No/kg	kg/ha
Stage I	1.0	1	1.0	18000	400.0	45	42	50.0	360
Stage II	3.0	1	3.0	6000	50.0	120	46	12.5	480
Stage III	3.0	3	9.0	2000	12.5	160	56	3.1	650

multiple stocking and harvesting in the same facility with a resultant range of size-groups present at any one time. This seems to give good results in pond systems where the different sizes of fish, perhaps consuming different components of the natural food, act as an intraspecific polyculture unit. In Taiwan milkfish farming the following stocking pattern per hectare, is typical: April (5000 fish at 5–150 g); May (2500 fish at 0.5 g); June (2500 fish at 0.06 g); July (2000 fish at 0.06 g); August (3000 fish at 0.06 g). Harvesting of 200–400 g fish takes place once a month from July to November. Unlike the May–August stockings those in April are from the previous year. Normally they would be larger by April but their growth is inhibited by keeping them in special wintering ponds at high densities, up to 500 000 ha^{-1}, and by using fish spawned late in the year.

In more intensive culture, a mixture of size-groups is normally avoided by frequent size segregation since it has been found in many species of fish (including bivalves) that a mixture of large and small individuals has adverse effects on the growth of the small ones. In some cases cannibalism will occur. Little attention is normally paid to the sex of the stock except where reproduction is either being encouraged or inhibited (§ 3.2), but in several species one of the sexes is faster growing. In these cases attention may be given to producing monosex lines.

The best growth results in all forms of aquaculture will be obtained with initial stock which is in good condition, unstressed, and free from disease. These areas are largely the responsibility of the hatchery/nursery operations, but it is often beneficial to have facilities for acclimation in situations where the early and on-growing environments are significantly different. Thus in the pond culture of fish and prawns, acclimation ponds (where the salinity can be easily controlled) or hapas (net enclosures) are often used.

It is possible to improve growth performance in aquaculture by the appropriate selection of organisms for stocking, but it does not follow that the growth advantage can be carried into the next generation by using the fast-growers as broodstock. Fast-growers do not necessarily beget fast-growers, and they may in any case be fast because of aggression in food competition rather than because of efficient food conversion. The latter is the desirable characteristic but is much more difficult to measure on an individual basis than overall growth performance. MOAV (1976) referred to fully controlled, replicated, large-scale, two-way selection experiments for five generations with common carp and concluded that mass selection for fast growth was not effective, although it was for slow growth. It was suggested that natural selection has probably already resulted in a pre-maturity growth plateau which cannot significantly be improved upon. However there may be some potential for artificial selection to improve upon the growth-rate after maturity is reached and therefore after the time at which most aquaculture stock is currently harvested.

Hybridization seems a much more useful genetic technique for manipulating growth-rate, and there are many examples of heterosis in both inter- and intraspecific hybrids. Much work on interspecific hybrids has been carried out on salmonids and tilapias and some results are shown in Table 6. Crossbreeding amongst Hungarian races of common

Table 6   Growth-rates of some salmonid hybrids. (After GJEDREM, T. (1976). *J. Fish. Res. Board Can.*, 33, 1094–9.) Mean body weights (g) at 11 months age.

| | | Female | | | |
		Char	Brown trout	Sea trout	Salmon
Male	Char	55.2	58.2	–	96.5
	Brown trout	73.3	41.8	–	7.7
	Sea trout	58.3	24.9	31.8	6.1
	Salmon	70.7	7.3	8.8	30.0

carp produced hybrids with 15–40% better growth than the pure lines, and they used 15–30% less food and had 2–5% less fat content.

Finally, work on several crustaceans has shown that eyestalk ablation can give enhanced growth-rate as well as encouraging maturation (§ 3.2). Thus it has been found that ablated lobster juveniles moulted more frequently than controls and showed a greater percentage weight gain per moult (Fig. 4–5). The teleost equivalent is to incorporate growth hormones into the diet (MATTY and CHEEMA, 1978; and § 4.4).

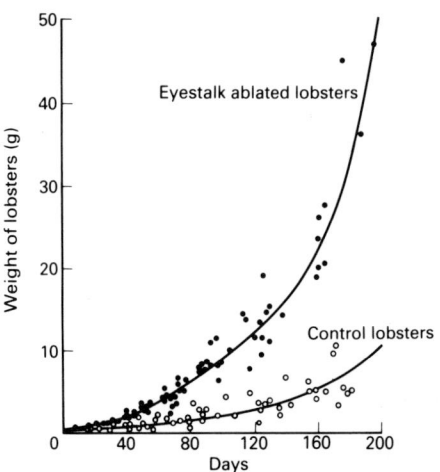

Fig. 4–5 The effect of eyestalk ablation on the growth of juvenile lobster, *Homarus americanus*. (After MAUVIOT, J. C. and CASTELL, J. D. (1976). *J. Fish. Res. Board Can.*, 33, 1922–9.)

4.5.2 *Environmental manipulations including heated effluents*

The environment in which the stock grows is mainly determined by site selection and type of holding facility. Since environmental factors can have a marked influence on growth-rate as well as survival (see § 5.5) an attempt will be made to optimize them through *site selection* (Table 7) and perhaps also through *manipulation of abiotic factors*. The most important factor to consider here is temperature, (*a*) because it has the most profound effects on growth and (*b*) because more information is available on the interaction between this factor and growth-rate; but other factors like photoperiod, salinity and dissolved oxygen level, also merit attention.

Each species has an optimum temperature for growth and conversion efficiency and ideally it would be desirable to provide the stock with a uniformly optimum temperature during the on-growing phase. Normally, however, temperature manipulation is prohibitively expensive. Thus for fattening turbot the optimum temperature is 14°C, which is only available for a part of the year in ambient seawater around U.K. coasts. To attain it in tanks would involve pumping water and heating it, and the heating would cost more than five times the value of the fish at the end of the on-growing period. The practical compromise may be to retain young turbot in on-shore tanks at 12–15°C over their first winter and then to grow them in ambient seawater. From egg to 500 g this process would take about 30 months, compared to 21 months in an optimum environment. On some Japanese eel farms the pond water is heated

Table 7 Percentage increase in weight of cupped oyster *Crassostrea gigas* (initial weight 7.2 g) under different densities and conditions in Strangford Lough, N. Ireland. (After PARSONS, J. (1974). *Aquaculture*, 3, 221–9.)

Position of oysters		% increase in mean weight after 6 weeks growth Initial numbers per 0.5 m^2		
		100	200	300
On raft		153	144	137
Sublittoral	on sea-bed	131	115	125
	on tray	139	130	153
Intertidal	on sea-bed	109	99	94
	on tray	118	103	104

during the winter months to maintain fast summer growth-rates and reduce the time taken to reach marketable size from 18 months to 12. Heating is by oil-fired boilers and is clearly expensive but some heat conservation is achieved by covering the ponds with large glass or plastic 'greenhouses'. Much more use could be made of covered ponds in temperate latitudes in fact, and work is now also beginning on the use of solar heating for raising water temperature. In Iceland, water from hot-water springs is used to rear young salmonids.

The main way of obtaining higher temperatures, however, involves the use of heated effluent from industry, in particular electricity generation (typically 8–10°C above ambient). Power stations in North America, Europe, the U.S.S.R. and Japan have associated fish farms. In the U.K. this applies to Ratcliffe-on-Soar, Ironbridge, Trawsfynydd, Hinckley Point, Wylfa, and Hunterston, but only 12 of the 137 U.K. power stations are considered suitable for commercial aquaculture. The species involved are eel, turbot, sole, common carp and rainbow trout, but it may become possible to farm tropical species like tilapia in future. The main biological advantage of using heated effluent is faster growth. Thus carp held in condenser water (mean temperature 25°C) at Ratcliffe-on-Soar power station were able to grow through the winter to reach 970 g in ten months, compared to only 180 g in ambient river water (mean 15.5°C) (ASTON *et al.*, 1976). Lobsters take about six years to reach 450 g in the wild, but only two years if maintained in heated effluent of 22°C.

A number of physical disadvantages are associated with the use of power station effluents, including the presence of chlorine which is used as a biocide to prevent fouling of the condensers. However, the chlorine is not always in concentrations which are toxic to fish, and the use of heat exchangers will avoid the problem if they are. In general, the advantages of heated effluent culture outweigh the disadvantages, and it will become increasingly important, particularly, but not exclusively, in temperate waters.

5 Minimizing Mortality

5.1 Patterns and sources of loss

Loss of stock is just as important as growth to the overall yield from a system, and commercial viability will be determined by the relative magnitude of these two processes. As long as growth exceeds loss then the living biomass will increase between initial stocking and harvesting. So a system incurring heavy losses may still be viable if the losses are more than matched by a high growth-rate of the survivors. Thus ranching of the chum salmon in Hokkaido, Japan, is a great success because the loss between release and recapture is *only* 97.8%, a level which would be disastrous in conventional salmonid culture. Even so, some loss will be acceptable in terms of economic viability in all forms of aquaculture, but there are strong incentives for increasing survival and certainly for preventing the occasional total loss to which all aquaculture is potentially susceptible.

In practice, losses may be due to escape, theft or 'natural' mortality. The first two can be minimized or prevented by appropriate construction of holding facilities, and the third includes all the sources of loss covered in this chapter. All can affect wild stocks, and are called natural because, in a fisheries biology context, they are thereby distinguished from fishing mortality (deaths caused by fishing). Fish in captivity are particularly susceptible to natural mortality for three reasons: (1) their artificial environment may be unsuitable because of inadequate understanding of requirements; (2) they cannot usually move away from any cause of potential mortality which would be the initial reaction of (mobile) wild fish; and (3), at high densities there will be a tendency for a natural regulation of numbers to occur through mortality (this may involve transmission of pathogens which is facilitated at high density). For these reasons, mortalities may be high during the initial culture of a species, but as its requirements become increasingly understood then good husbandry and domestication should provide conditions concomitant with high survival and good growth.

Abiotic factors, diet, predators and *pathogens* are the main mortality factors, but because of interactions between factors, it is not always possible to determine a primary cause of mortality. Thus an unfavourable environment or diet may increase susceptibility to disease which may in turn increase the chance of predation. A specific example links copper ($30-60$ mg l^{-1}) in the water with vibriosis disease of eels. Thus mortality control might involve attention to water quality, construction materials and diet, in addition to a direct attack on other organisms or enhancement of the fish's own defence mechanisms. The four sets of mortality factors can all induce stress without directly causing mortality (if wounding is considered as a mild form of predation). Stress has been equated with unhappiness, but it is also referred to as the sum of all physiological responses by which an animal endeavours to re-establish homeostasis in the face of disrupting physical or chemical forces. The responses are energy-consumptive and a stressed animal is unlikely to grow. Further, its probability of death from any mortality factor is increased.

Another problem which makes it difficult to analyse mortality causes is the limited amount of data usually available on the extent and pattern of the mortality. This requires accurate counting of stock at stocking and harvesting and, if possible, at intervals in between, but very often such large numbers of small animals are involved that the procedure is very time-consuming. Further, frequent counting may itself induce stress. The alternative is to count dead bodies, but whereas these may be quite conspicuous in intensive cage, raceway or tank culture, they will not be in other forms of aquaculture. However, if some fresh corpses or preferably still living, but moribund, fish are available, observations on behaviour and appearance (including colouration), and an examination of the fish may enable the cause to be diagnosed. In any case, dead and dying fish, which are potential sources of infection if pathogens are involved, are usually removed wherever possible and buried in lime-pits. A procedure for post-mortem examination of salmonids is given by ROBERTS and SHEPHERD (1974).

5.2   Abiotic factors, nutrition and predation

5.2.1   *Abiotic factors*

All organisms have lethal limits in response to abiotic factors such as temperature, salinity, pH, dissolved gases, heavy metals and other chemical factors. As they affect the fish, these are water-quality criteria but it is important to remember that the quality of the influent water is determined by geological, edaphic, climatic and human factors often a considerable distance away from the farm site. It can be further affected by the materials used in the construction of holding facilities (including piping and earth pond soils) and by the activities of fish, farmer and other organisms on site. This emphasizes the importance of careful site selection, monitoring, and maintenance of adequate water quality.

Although the different abiotic factors can have a range of physiological effects, a major reason why fish die is that they are unable to obtain enough oxygen. This may be because the dissolved oxygen concentration is too low, because the gills are damaged, or because the haemoglobin-uptake mechanism is impaired. Death can of course be rapid and aversion of a disaster even if the farmer is lucky enough to have prior warning, is only possible by prompt action. Whether the oxygen problem is caused by environmental or physiological factors, some relief is usually possible by increasing dissolved oxygen levels (§ 2.2).

Dissolved oxygen levels are affected by movements of wind and water, water temperature, and the biomass and activity of the stock and other organisms (especially plants). Increases in biological oxygen demand can arise if plankton blooms occur, but more commonly, it is due to bacterial populations increasing in response to an accumulation of faecal matter or uneaten food. Other aspects of self-pollution can reduce the efficiency of oxygen uptake by the fish. These are primarily the effects of *unionized ammonia*, *nitrite* and *carbon dioxide*.

Ammonia, the main excretory product of aquatic animals, and CO_2, both affect pH in water which is not adequately buffered. Further, together with temperature, pH can affect the proportion of total ammonia which is in the toxic, unionized form (UIA) and here the lower the pH the better; for example, for a total ammonia concentrate of 12 mg l^{-1}, pH 6.5, temperature 10°C, the UIA is 0.007 mg l^{-1}, whereas at pH 8.2 and temperature 15°C, the UIA is 0.499 mg l^{-1}. The former concentration is non-toxic, the latter acutely lethal to rainbow trout (FORSTER *et al.*, 1977). UIA is toxic because it easily passes through gill membranes and up to 25 times the environmental concentration may be present in the fish. High concentrations of ammonia in the blood causes hyperplasia, proliferation, clubbing and consolidation of the gill lamellae leading to stress and eventually to mortality. Mortalities of salmonid fry due to blue-sac disease have also been ascribed to high concentrations of ammonia. Oxygenation helps by accelerating the natural conversion of ammonia to nitrite by bacteria. However, nitrite itself can be toxic because it converts haemoglobin to methaemoglobin which is unable to combine reversibly with oxygen. Levels of 0.14–0.55 mg NO_2-N^{-1} have been quoted as lethal for salmonids, but penaeid prawns and turbot seem to be more tolerant. Nitrite in water is converted to nitrate which is non-toxic, at least at the concentrations likely to be encountered.

The extent and accumulation of free CO_2 in water is largely dependent on the buffering capacity of the water. There is little information on its toxicity but prolonged exposure of trout to sub-lethal free CO_2 concentrations may sometimes result in a kidney disease known as nephrocalcinosis.

Assuming that there are optimal values for each abiotic factor, then departure from the optima will produce a variety of effects depending on the degree of departure, the duration of exposure, other environmental factors and conditions of the stock. The effects will vary from a slight inhibition of growth in some individuals to a sudden total mortality. Economic criteria will largely determine at what point along this line the situation is serious and merits treatment.

Rapid changes in the abiotic environment are much more likely to cause problems than gradual changes and here the ability of the organisms to acclimate is an important consideration. The transfer of sea-farmed salmonids from their freshwater hatchery/nursery environment is potentially a period of high mortality brought about by farm operations and the need to restrict the period of high-cost freshwater phase to a minimum. Here the young fish must be preadapted in terms of their osmoregulatory mechanisms. For rainbow trout, although transfer may take place at any time of year, a sudden change from fresh to salt water with good survival is only possible if the fish are above 60 g in weight. However, it has been found that if water of intermediate salinity can be used to acclimate the fish, the transfer can take place at weights as low as 10 g. An alternative method is to feed a high salt diet (10% salt) for two weeks before transfer.

5.2.2 Nutrition

Nutrition as a mortality factor rather than a growth inhibitor will occur when deficiencies, excesses and toxicities in the diet are extreme or if they persist over a long period. An example of a nutritional disease is liver lipoid disease in rainbow trout which is associated with high lipid or carbohydrate levels and can cause up to 75% mortalities.

Studies on channel catfish have shown that the first symptoms of vitamin deficiencies tend to be depressed appetite and reduced growth, followed by discolouration, lack of coordination, nervousness, haemorrhage, lesions, fatty livers and susceptibility to bacterial infections. Ultimately death may occur.

As the nutritional requirements for optimal growth become better understood and as long as artificial diets approximate to these requirements, then nutritional problems are not likely to be important causes of mortality in aquaculture.

5.2.3 Predators

Predation represents the main cause of mortality in wild populations and this will also be the case in ranching and stocking. Particularly in inland waters, there may therefore be some attempt to control the populations of fish and bird predators. In European bivalve culture predation by crabs (Carcinus maenas), starfish (Asterias rubens) and oyster drills (Urosalpinx cinerea) can cause serious losses and is most important in bottom culture, least in raft culture. Starfish can be removed from beds by applying quicklime or by towing across the beds fibrous material to which the animals adhere. Crabs are much more difficult to control, but larger ones can be excluded by netting enclosures or containers. Failing this, it is possible either to use off-bottom culture methods, or to site farms away from areas where crab populations are high.

Predatory fish can enter fish ponds either in the influent water supply or as fry amongst the fry of the cultured species if these are harvested from the wild. Screens at the point of water flow into the ponds will of course prevent the entry of large predatory fish but the young stages will still be able to pass through. Thus in milkfish ponds, tarpon (Megalops cyprinoides) and ten-pounder (Elops hawaiiensis) enter in both ways and can prey on the milkfish fry and fingerlings. The normal process of drying the ponds before stocking will periodically remove or kill most undesirable organisms, but if this is not possible then tea-seed meal (containing saponin) or other nonpersistent poisons can be applied before stocking.

At least if predatory fish do occur in ponds they can be harvested as a secondary crop, so that mortality of the prime species does not constitute a total loss of energy to the system. If predatory birds or mammals are involved, however, this is unlikely to apply. Herons (*Ardea cineraria*) and mink (*Mustela vison*) are two important predators causing losses in U.K. trout farms.

A final type of predation which deserves special mention is cannibalism. The decapod crustacea include several species (e.g. caridean prawns, lobsters and crayfish) which are cannibalistic if kept at high densities without refuges, and it can be a serious constraint to the development of intensive aquaculture for such species. In situations where the life history is completed in captivity, the trait of cannibalism (assuming there is variability in the aggressive tendency and that some of this has an additive genetic basis) may become increasingly prevalent as only the most aggressive will survive and from these the broodstock will be selected. Conversely it may be possible to select for non-aggression when rearing and choosing broodstock as a way of reducing the problem. Another approach is based on the fact that adult females immediately after moulting are mated with rather than eaten by the males, and this protection is probably achieved by secreted pheromones. It is thought that isolation of the pheromone and its application in intensive culture units could reduce cannibalism.

5.3 Parasites and fungi

Wild fish populations have an associated parasite fauna whose members have rarely been regarded as a direct cause of mortality. In those cases where heavy losses do occur it is possible that human factors (such as pollution) have stressed the populations to the extent that their susceptibility to infection is increased. An equilibrium naturally maintained between host and parasite may also be disturbed by the introduction of a new host or parasite strain. What are normally considered unimportant parasites, therefore, may under suitable conditions become much more serious, even fatal. Aquaculture is capable of providing such conditions and several examples are given by MCVICAR and MACKENZIE (1977).

It has been said that all parasites are lethal to their hosts if they are present in large enough numbers, so in looking at parasitism as a cause of mortality, one is concerned with the numbers of hosts infected and the numbers of parasites per host. Many factors affect the level of infestation and the situation is a complex one involving the population dynamics of parasite and often more than one host species. However, some relevant data on the cestode *Caryophyllaeus laticeps* in bream *Abramis brama*, are provided by ANDERSON (1974). This parasite is found as an adult in bream and as larvae in the tubificid worms on which the bream feeds. Numbers vary seasonally and with the age of the host, and equilibrium is maintained in the natural environment as a result of over-dispersion of the parasites and a periodicity in immigration of larvae into the adult population due to seasonal feeding pattern of the fish hosts. There is also a periodicity in death-rate within the fish tissues probably due to a temperature-dependent immune response. Such an equilibrium is more likely to be disrupted in an aquaculture environment, although in the most intensive culture where intermediate hosts are unlikely to be present and the diet is artificial, parasites like *C. laticeps* would not be so important as those with direct life-cycles.

Ichthyophthirius multifiliis, a ciliate protozoon, falls into the latter category and it causes white-spot or ichthyophthiriasis which can be a devastating disease of cultured freshwater fish. It seems to affect all teleost species (although it is only endemic to cyprinids) in all climates. It has been studied by HINES and SPIRA (1974) in a series of papers of which the last is given as a reference. The mirror strain of common carp was used as host, and this is a fish affected by the disease in Israeli fish ponds. The parasite invades the epithelium of skin and gills and disturbs respiratory and excretory processes. When the carp were experimentally infected with 40 mature parasites they recovered after 18 days of mild disease symptoms, but 400 parasites per fish always caused death after 22–25 days. Fish recovering from intermediate, but still heavy, infestation acquired immunity to the normally lethal levels for at least eight months. The parasites which swim freely in dilutions of normal serum are immoblized by serum from recovered carp, but the effective barrier to reinfection operates at the level of the external mucus covering where serum proteins are also present. (The skin and its associated mucus is an important site of fish defence mechanism against many pathogens.) The larvae grow within the epidermis of the fish and emerge as a 1 mm adult which encysts on the pond bottom before releasing up to 500 small infection stages.

The main groups of parasites of actual or potential importance in aquaculture are considered below, together with methods of control.

1   *Protozoa*. Apart from ichthyophthiriasis several other important diseases are caused by protozoa: one example is the whirling disease of trout caused by *Myxosoma cerebralis*. It is avoided by not growing young fish in earth ponds, since the parasite must spend up to six months in the mud before being able to infect, and then it is only young fish (6–8 cm). Once ingested it invades the gut wall and passes into the skull and spine cartilage, causing chronic debility and sometimes high mortalities. Haplosporidians (e.g. *Minchinia nelsoni*) cause disease amongst oysters.

2   *Helminths*. There are some important *cestodes* affecting both fish and human consumers, especially where the fish is the intermediate host to the

pleurocercoid larvae living in the muscle tissue. Parasitic *nematodes* are essentially a marine group, and larval anisakidids could cause disease in farms where marine fish offal is used as food. As well as debilitating the stock, their presence in the musculature could affect marketability, but so far they do not seem to have been a problem.

Amongst the *monogeneans*, species of *Dactylogyrus* are common on the gills of salmonids and may damage them, and *D. vastator* is sometimes a real menace to young fish in warm water ponds, able to cause very high mortalities within a few days. Two species of monogenean are listed among the principal causes of mortality in Japanese yellowtail culture, and it is tempting to conclude that this is the most important group of helminths in aquaculture.

The *digeneans* are usually less of a problem because their life-cycle requires the presence of fish and two other hosts, one a mollusc. MCVICAR and MACKENZIE (1977), however, have discussed the dangers of polyculture in this context.

3 *Crustacea.* Apart from species of *Argulus*, or fish lice, which can cause heavy mortalities, particularly in freshwater fry ponds in the tropics, the main crustacean parasites are copepods. Thus gill-maggot (*Salmincola edwardsii*) and salmon louse (*Lepeophtheirus salmonis*) are salmonid examples which can become epidemic in fish farms. The salmon louse can be a serious problem in salmon farms in Norway and Scotland: the parasite rasps away at the head and opens the way for secondary infections to take place.

The copepod *Mytilicola intestinalis* is a common parasite of the mussel particularly in bottom culture, but is more likely to affect growth and condition than survival.

4 *Fungi.* Fungi are characteristically a secondary infection of fish although the well-known *Saprolegnia parasitica* will attack healthy eggs as it spreads from dead ones. Other serious fungal parasites are *Aphanomyces astaci* causing crayfish plague, *Labyrinthomyxa marina* causing shell disease of oysters, and *Pythium* spp. affecting cultured red algae like *Porphyra* in Japan.

5 *Control methods.* The best prophylactic techniques against parasite problems in aquaculture are *hygiene* and *good husbandry*, i.e. the appropriate water quality, food and stocking density for the species concerned should be maintained; dead and dying fish, uneaten food and waste products should not be allowed to accumulate; and conditions necessary for parasite life-cycle completion, such as mud for *Myxosoma cerebralis* and molluscs for digeneans, should be avoided. The use of bore-hole water, filters and formulated diets also helps.

Chemotherapeutic treatment has now been developed which enables the prevention or successful control of losses due to most ectoparasites without harming the host. Some of the main chemicals in use are copper sulphate, formalin, chloramine-T, malachite green and masoten. The latter is an organophosphorous compound particularly effective against crustaceans. In addition some success has been achieved by oral administration (in the food) of di-n-butyl tin oxide for certain gut parasites and neguvan (trichlorfon) for salmon lice.

Further details on diseases relevant to aquaculture are given, for example, by SARIG (1976), SINDERMAN (1970), and ROBERTS and SHEPHERD (1974).

5.4 Bacteria and viruses

5.4.1 Bacteria

Not all the bacteria associated with diseased fish are true pathogens. Many are secondary invaders and in terms of disease control it is important to distinguish the two. Like viruses, bacteria may precipitate severe losses, but the onset is usually less sudden. The fish typically go off their feed one or two days before they start to die.

An important disease of salmonids (and now also of a variety of other freshwater fish and at least one marine species) is furunculosis, caused by the bacterium *Aeromonas salmonicida*, an obligate pathogen. High mortalities may occur in market-sized fish. Damage is caused by toxins and, in affected fish, the spleen enlarges and darkens, white cells in the blood are scarce and there may also be muscle lesions. Work done at the MAFF diseases laboratory at Weymouth has shown that some fish are carriers (one farm has 100% carriers but no disease), and fish which have died from the disease produce enough bacteria to infect healthy ones. The bacteria can survive without hosts for 26 days in brackish water, 18 days in freshwater and 8 days in seawater, and for 14 days in mucus and mud on dried nets. They do not readily transmit vertically, however, as survival on the egg is poor. These are important considerations in the selection and application of control measures.

The myxobacteria group are responsible for a number of well-known fish diseases such as fin-rot, tail-rot, bacterial gill disease, bacterial kidney disease and columnaris.

Vibrio anguillarum causes vibriosis disease in fish through the release of toxins acting on the blood cells, causing haemorrhagic septicaemia when acute, dermal petechial haemorrhage and bloody abscesses and ulcerations when sub-acute or chronic. Once established, 95–98% of an infected population can die within a few days.

A bacterial disease affecting lobsters is gaffkemia caused by *Aerococcus viridans* var. *homari*. Lobsters seem to have very little resistance to this pathogen which enters through wounds and affects the hepatopancreas. Other bacteria adversely affecting crustaceans are chitoclastic species which secondarily infect scratched or damaged exoskeletons.

Control of bacterial infections can be by *chemotherapy*. Furanace has a broad spectrum of activity, and is added to water, usually in a bath. Some success against gaffkemia has been achieved in this way with vancomycin. Alternatively drugs such as sulphonamides and nitrofurans may be added to the feed. However, difficulties with this type of administration are that the feeding rate is critical in dose calculations and the fish must be willing to feed. There is also some concern over the development of drug resistance by fish bacteria and the transfer of this resistance to strains infecting man.

To some extent these difficulties can be overcome by using *antigens* (vaccines) injected into the fish. Although this can be time-consuming, it is now the preferred method of vibriosis control among commercial salmon farmers on the Pacific coast of North America. An improvement in the administration of antigens has been the use of *hyperosmotic infiltration* which is less time-consuming and causes a lower treatment mortality. The fish to be treated are placed initially in a high concentration of NaCl or urea, and then in an osmotically weak solution containing the antigen which is carried across the gills into the blood by osmosis. (ANTIPA and AMEND, 1977.)

As with parasites, however, the best approach to bacterial disease is prevention through reduction of stress by attending to hygiene and husbandry requirements.

5.4.2 Viruses

One of the most important virus diseases in salmonid culture is *infectious pancreatic necrosis* (IPN) which mainly attacks fry and fingerlings. It was the first fish virus isolated (1957). Although endemic in North America, Japan and Europe (including the U.K.), many farms using spring-water are free of disease and these can become valuable sources of eggs. Eggs can be, and are, routinely disinfected by the application of iodophores (e.g. 5 min at 30 ppm of active iodine) which kill the virus. In the absence of hosts, the particles also lose 99% of their infectivity if kept at 4°C for 10–12 weeks in fresh water.

Viral haemorrhagic septicaemia (VHS) is not so contagious as IPN and is a disease of rainbow trout fingerlings restricted so far to mainland Europe. *Infectious haematopoetic necrosis* (IHN) is confined to North America and Japan.

Not only salmonids of course are affected by acute virus disease, but it is the widespread intensive culture of these fish which has perhaps resulted in such serious problems. In the same way, *channel catfish virus* (CCV), and *spring viraemia of carp* (SVC) occur in intensive culture in the southern U.S.A. and Europe respectively. SVC is regarded as the viral form of infectious dropsy of carp, and although at present absent from the U.K. the other form, erythrodermatitis, is fairly common. No aetiological organism has yet been positively identified for erythrodermatitis, and this also applies to many of the tumours found in fish, and to UDN (ulcerative dermal necrosis) which has seriously affected wild salmon in the U.K. but not those under culture.

Not all virus diseases, therefore, are associated with intensive aquaculture, but they will almost certainly become more important as the industry expands and intensifies. Many currently unexplained mortalities in cultured teleosts, molluscs and crustacea probably involve viruses.

5.5   Manipulations of stock and environment

The appropriate selection of species, source of stock, stocking density and the prompt removal of dead and moribund fish are the usual manipulations of stock which will help to reduce mortality on fish farms. To these may be added the environmental manipulations associated with site selection, construction of holding facilities, diet, and the maintenance of an adequate water quality. These have been elaborated in previous sections but some further discussion is needed of control of fish movements, genetics in relation to disease control, and the scope of environmental manipulation.

5.5.1   *Restrictions on fish movements*

The pathogens associated with fish diseases often have a restricted range, being endemic to certain areas of the world, and within those areas being restricted perhaps to particular water courses, species or strains of fish. Harmonious relationships between host and pathogen are often disturbed not only by growing fish at high densities in suboptimal conditions, but by introducing species of fish and their associated pathogens to areas where they are not indigenous. The deleterious effect is normally of exotic pathogens on indigenous hosts, and many of today's disease problems arise from uncontrolled introductions in the past.

As a result there is concern about world-wide trade in live fish and eggs, and in many countries there is legislation to control the importation and internal movement of teleost and bivalve (but not usually crustacean) species. In some cases this is vital for the control of infections, but it will also initially be a constraint to realizing the full productivity of particular situations, and ineffective because of the inevitable loopholes.

In 1975, new licensing conditions of the Disease of Fish Act, 1937, which apply to the importation of freshwater teleosts and their eggs, were introduced in the U.K. Importation of live salmonid fish remains prohibited under the Act, and elsewhere, import licences must be accompanied by health certificates showing that samples of the farm stock have been tested over a two-year period at intervals of six months and found free of IPN, VHS, IHN and dropsy. The main loophole in the legislation concerns ornamental fish which often find their way into natural waters and upon which there is no effective restriction on imports. There is no control over the movements of crustacea but the Molluscan Shellfish (control of deposit) Order, 1974, limits the import and movement of marine shellfish in England and Wales, in order to control (mainly oyster) pests and diseases.

Movement control is best carried out at source and needs to have an international basis. Accordingly the International Office of Epizootics, the European Inland Fisheries Advisory Committee (EIFAC) and FAO are in the process of formulating an international convention for the control of selected communicable diseases, based on inspection at source combined with certification.

5.5.2   *Genetics and disease resistance*

Another approach to disease control involves the use of genetic manipulations. Although little work has been done so far, some progress has been made in the artificial selection for dropsy resistance in carp and presumably some low level natural selection is taking place anyway. Experiments on common carp suggest that susceptibility to epidermal epithelioma and swimbladder inflammation is controlled by recessive genetic factors. The Virginia Institute of Marine Sciences has developed strains of oysters which are disease-resistant, by using as parents the survivors of a mass mortality in the mouth of Chesapeake Bay in which most deaths were due to the haplosporidian *Minchinia nelsoni*.

5.5.3   *Environmental manipulation*

The environmental manipulations designed to optimize growth (§ 4.5) will at the same time tend to optimize survival since a rapidly growing fish is likely to be healthy and free from stress. However, specific manipulations, which may or may not be at variance with the requirements for rapid growth, can be used to increase resistance or to reduce the immigration and survival of pathogens. Thus the use of low salinity areas for on-growing oysters may suppress mortality caused by the fungus *Labyrinthomyxa marina*. The substitution of concrete or fibreglass tanks for mud-ponds in trout fry rearing will prevent myxosomatosis, although air-drying of the pond mud can also be effective in killing the protozoan spores.

In more general terms, the use of borehole water, compounded diets, gravel filters and disinfectant water baths in hatcheries, burying dead fish in lime pits, discouraging birds and draining and drying earth ponds for at least a month each year will help to keep salmonid stocks free from serious disease. Maintenance of high dissolved oxygen levels, a pH between 6.5 and 9.0, and provision of a suitable diet will help to avoid stressing the fish and lowering their resistance. There may also be other requirements, in particular those related to social behaviour, which are being ignored in current aquaculture practice. Nothing seems to be known of such requirements in aquatic organisms, but their importance in domestic birds and mammals has been exposed by KILEY-WORTHINGTON (1977) and could promise to be a ripe field for research.

Appendix

Some of the main species used in modern aquaculture, indicating countries where cultured, type of culture (I–VIII as in Table 1), and whether fresh water (F), brackish water (B), or marine (S).

PHYLUM MOLLUSCA – Class *Bivalvia*

Species	Common name	Countries	Type	Water
Mytilus edulis	common mussel	Spain, Netherlands, France, etc.	I, II	S B
Mytilus smaragdinus	green mussel	Philippines, Thailand, etc.	II	S B
Perna perna	mussel	Venezuela, New Zealand	II	S
Anadara granosa	cockle	Malaysia, Thailand, etc.	I	S B
Crassostrea gigas	Pacific oyster, cupped oyster	Japan, Europe, U.S.A.	I, II, V, VI	S B
(several other species inc. *C. virginica, C. rhizophorae*)				
Ostrea edulis, O. virginica	Flat oyster	Europe, U.S.A.	I, II, V, VI	S B
Patinopecten yessoensis	scallop	Japan	II	S
Pecten maximus	great scallop	U.K.	II	S

PHYLUM MOLLUSCA – Class *Gastropoda*

Species	Common name	Countries	Type	Water
Haliotis discus	ormer, abalone	Japan	V, VIII	S

PHYLUM ARTHROPODA – Class *Crustacea* (Order Decapoda)

Species	Common name	Countries	Type	Water
Penaeus monodon	jumbo tiger prawn	Philippines, Indonesia, etc.	II, IV	S B
Penaeus japonicus	kuruma prawn	Japan	V, VIII	S
(several other penaeid spp. inc. *P. indicus, Metapenaeus monoceros*)				
Macrobrachium rosenbergii	giant freshwater prawn	S. E. Asia, U.S.A., etc.	VI, VIII	B F
Homarus gammarus, H. americanus	lobsters	Europe, U.S.A. (experimental)	V, VIII	S
Pacifastacus leniusculus	signal crayfish	Europe	V, VIII	F
Procambrus clarkii	red swamp crayfish	southern U.S.A.	VI?	F

PHYLUM CHORDATA – Class *Osteichthyes*

Order Acipenseriformes

Species	Common name	Countries	Type	Water
Acipenser stellatus	sturgeon	U.S.S.R.	V	B F

Order Gonorhynchiformes

Species	Common name	Countries	Type	Water
Chanos chanos	milkfish	Philippines, Indonesia, Taiwan	II, IV	B F S

Order Salmoniformes

Species	Common name	Countries	Type	Water
Onchorhynchus kisutch	coho salmon	U.S.A., Canada, France, etc.	V, VIII	F + F S
Onchorhynchus keta	chum salmon	U.S.A., Canada, Japan	V, VIII	F + F S

Species	Common name	Distribution		
Salmo salar	Atlantic salmon	Norway, Scotland, Canada	V, VIII	F + F S
Salmo trutta	brown trout	World-wide } including tropical	V, VIII	F + F S
Salmo gairdneri	rainbow trout	World-wide } highland areas	V, VIII	F + F S
Salvelinus fontinalis	American brook charr	N. America, Europe	V, VIII	F
Plecoglossus altivelis	ayu	Japan	V, VIII	F
Coregonus peled, etc.	whitefish (several spp)	U.S.S.R.	V, VIII	F
Order Anguilliformes				
Anguilla anguilla, A. japonica	eel	Europe, Japan	II, IV	F B S
Order Siluriformes				
Ictalurus punctatus	channel catfish	U.S.A.	VI, VIII	F
Clarias batrachus	walking catfish	S.E. Asia, India	VI, VIII	F
Order Cypriniformes				
Carassius carassius	crucian carp	Europe, Asia	VI, VIII	F
Cyprinus carpio	common carp	World-wide	VI, VIII	F
Ctenopharyngodon idella	grass carp } Chinese	Europe, Asia, U.S.A.	VI, VIII, IV	F
Hypophthalmichthys molitrix	silver carp } carps	Asia, Israel	VI, VIII, IV	F
Labeo rohita	rohu	India	VI, VIII, IV	F
Catla catla	catla } Indian major carps	India	VI, VIII, IV	F
Cirrhina mrigal	mrigal }	India	VI, VIII, IV	F
Order Perciformes				
Lates calcifer	sea-bass, cock-up	S.E. Asia	II, IV, VIII	B S
Dicentrarchus labrax	bass	Mediterranean	II	B S
Mugil cephalus	grey mullet	Israel, Hawaii, S.E. Asia	II, IV	B S
(several other species of grey mullet, Mugilidae)				
Trichogaster pectoralis	gourami	S.E. Asia	VI, VIII	F
Siganus canaliculatus	rabbitfish	Philippines	II, IV	S B
Seriola quinqueradiata	yellowtail	Japan	IV	S
Sarotherodon mossambica	tilapia	World-wide in tropics	VI, VIII	F B
(several other species of tilapia of the genera *Sarotherodon* and *Tilapia*)				
Boleophthalmus chinensis	mud-skipper	Taiwan	IV	S B
Order Pleuronectiformes				
Scophthalmus maximus	turbot	U.K., France	VIII	S
Solea solea	Dover sole	U.K., Belgium	VIII	S
Order Tetraodontiformes				
Fugu rupripes	puffer	Japan	VIII	S
ALGAE				
Porphyra yezoensis (Rhodophyceae)	nori	Japan, China	VI	S B
(several other species of Rhodophyceae, Phaeophyceae, Chlorophyceae)				

Bibliography

ANDERSON, R. M. (1974). *J. Anim. Ecol.*, **43**, 305–22.

ANON (1977). Freshwater fish farming in China – report by an FAO mission. *Fish Farming International*, **4**(3), 18–23.

ANTIPA, R. and AMEND, D. F. (1977). *J. Fish. Res. Bd Can.*, **34**, 203–8.

ASTON, R. J., BROWN, D. J. A. and MILNER, A. G. P. (1976). *Fish Farming International*, **3**(2), 41–4.

*BARDACH, J. E., RYTHER, J. H. and McLARNEY, W. O. (1972). *Aquaculture: the Farming and Husbandry of Freshwater and Marine Organisms*, Wiley, New York.

CHAUDHURI, H. (1976). *J. Fish. Res. Bd Can.*, **33**, 940–7.

CHEN, T. P. (1976). *Aquaculture Practices in Taiwan*, Fishing News Books, Farnham, Surrey.

CUSHING, D. H. (1977). *Science and the Fisheries*. Studies in Biology no. 85. Edward Arnold, London.

EDWARDSON, W. (1976). *Fish Farming International*, **3**(4), 10–13.

ELLIOTT, J. M. (1976). *J. Anim. Ecol.*, **45**, 923–48.

FORSTER, J. R. M. *et al.* (1977). *Fish Farming International*, **4**(1), 10–13.

HALVER, J. E. (1976). *F.A.O. Technical Conference on Aquaculture*, Paper R. 31. In: PILLAY (1978).

HICKLING, C. F. (1971). *Fish Culture*, revised edition. Faber, London.

HINES, R. S. and SPIRA, D. T. (1974). *J. Fish Biol.*, **6**, 373–8.

HUET, M. (1973). *Textbook of Fish Culture, Breeding and Cultivation of Fish*, 4th edition. Fishing News Books, Farnham, Surrey.

JONES, A. (1972). *Lab. Leaflet Fish. Lab. Lowestoft*, **24**.

KERR, N. M. (1976). *Proc. Roy. Soc. Edinburgh, Ser. B.*, **75**(4), 263–70.

KILEY-WORTHINGTON, M. (1977). *Behavioural Problems of Farm Animals*. Oriel, Stocksfield.

KORRINGA, P. (1976). *Farming Marine Organisms Low in the Food Chain*. Elsevier, Amsterdam.

KUO, C-M., NASH, C. E. and SHEHADEH, Z. H. (1974). *Aquaculture*, **3**, 1–14.

LOOSANOFF, V. L. and DAVIS, H. C. (1963). *Adv. Mar. Biol.*, **1**, 2–136.

MCVICAR, A. H. and MACKENZIE, K. (1977). In: *Origins of Pest, Parasite, Disease and Weed Problems*, edited by J. M. CHERETT and G. R. SAGAR, Blackwell, Oxford, 163–82.

MATTY, A. J. and CHEEMA, I. R. (1978). *Aquaculture*, **14**, 163–78.

MILNE, P. H. (1972). *Fish and Shellfish Farming in Coastal Waters*. Fishing News Books, Farnham, Surrey.

MILNE, P. H. (1975). *Fish Farming International*, **2**(3), 15–19 and **2**(4), 18–21.

MOAV, R. (1976). *F.A.O. Technical Conference on Aquaculture*, Paper R.9. In: PILLAY (1978).

MOAV, R., WOHLFARTH, G., SHROEDER, G. L., HULATA, G. and BARASH, H. (1977). *Aquaculture*, **10**, 25–43.

NASH, C. E. and KUO, C-M. (1975). *Aquaculture*, **5**, 119–33.

NEW, M. B. (1976). *Aquaculture*, **9**, 101–44.

ODUM, W. E. (1974). *Environm. Conserv.*, **1**, 225–30.

O'SULLIVAN, B. W. and WILSON, J. H. (1976). *Fish Farming International*, 3(3), 12–15.

PASTAKIA, C. M. R. (Editor) (1978). *Report of the Proceedings of a Conference on Fish Farming and Wastes, January 1978*. Janssen, London. 151 pp.

*PILLAY, T. V. R. (editor) (1978). *Advances in Aquaculture*. Fishing News Books, Farnham, Surrey.

PITCHER, T. J. (1976). *Environm. Conserv.*, 4, 59–65.

PURDOM, C. F. (1972). *Lab. Leaflet Fish. Lab. Lowestoft*, 25.

ROBERTS, R. J. and SHEPHERD, C. J. (1974). *Handbook of Trout and Salmon Diseases*, Fishing News Books, Farnham, Surrey.

ROSENTHAL, H. (1976). *F.A.O. Technical Conference on Aquaculture*, Paper E. 67. In: PILLAY (1978).

SARIG, S. (1976). *F.A.O. Technical Conference on Aquaculture*, Paper R. 32. In: PILLAY (1978).

SHELBOURNE, J. E. (1975). *Fish. Invest. Lond.*, Ser. 2, 27(9).

SINDERMANN, C. J. (1970). *Principal Diseases of Marine Fish and Shellfish*. Academic Press, New York and London.

SORGELOOS, P. and PERSONNE, G. (1975). *Aquaculture*, 6, 303–17.

WEATHERLEY, A. H. (1972). *Growth and Ecology of Fish Populations*. Academic Press, London and New York.

WEATHERLEY, A. H. (1976). *J. Fish. Res. Bd Can.*, 33, 1046–58.

WEBBER, H. H. and RIORDAN, P. F. (1976). *Aquaculture*, 7, 107–24.

WICKINS, J. F. (1976). *Oceanog. Mar. Biol. Ann. Rev.*, 14, 435–508.

* These books, together with the periodicals *Aquaculture*, *Fish Farming International*, and *F.A.O. Aquaculture Bulletin*, are particularly recommended as sources of information on aquaculture.